More praise for *Adam's Curse*

"Conversational, often witty and chock-full of real-world examples, *Adam's Curse* deftly draws threads from history, anthropology, biochemistry, geology, sociology and population genetics. . . . Sykes' writing style brings to mind the clarity of Steven Jay Gould and the wonder, bemusement and gentle chiding of Carl Sagan."
—John Mangels, *Cleveland Plain Dealer*

"Sykes's book makes eye-opening reading for both sexes."
—*Sunday Times* (London)

"Sykes is an engaging writer and largely succeeds in leading a general-interest reader through the thickets of how cellular genetics works. It's no simple task—and in the end, the reader emerges with new ways of thinking about an old cliché: Indeed love is a battlefield."
—Scott Thomas, *Buffalo News*

"Like a cross between a detective thriller and a somewhat demanding course in Genetics 101 taught by a charming, entertaining teacher, *Adam's Curse* guides us through the developments and discoveries that contributed to what we now know about the mechanism of biological inheritance."
—Francine Prose, *O Magazine*

"Well-known Oxford geneticist Sykes, in this lively and thought-provoking book, gives a genetic twist to the battle between the sexes."
—*Publishers Weekly*

"Sykes takes the reader on a fascinating journey with no clear destination and only the most uncertain outcome, and all the while he is funny, provocative and smart. *Adam's Curse* is a great waltz with the intricacies of the male and female stripped down to their biological essences."
—Retha Oliver, *San Antonio Express News*

A FUTURE WITHOUT MEN

ADAM'S CURSE

BRYAN SYKES

W. W. NORTON & COMPANY

NEW YORK LONDON

For information about permission to reproduce selections from
this book, write to Permissions, W. W. Norton & Company, Inc.,
500 Fifth Avenue, New York, NY 10110

Manufacturing by The Haddon Craftsmen, Inc.
Production manager: Amanda Morrison

Library of Congress Cataloging-in-Publication Data

Sykes, Bryan.
Adam's curse : a future without men / Bryan Sykes.—1sr American ed.
p. ; cm.
Includes index.
Originally published: London : Bantam Press, 2003.
ISBN 0-393-05896-4
1. Y chromosome—Popular works. 2. Sex (Biology)—Popular works.
3. Human evolution—Popular works.
[DNLM: 1. Genetics—Popular Works. 2. Sociobiology—Popular Works.
3. Y Chromosome—Popular Works. QH 447 S983a 2004] I. Title.
QH600.5.S98 2004
599.93'6—dc 22
2004003628

ISBN 0-393-32680-2 pbk.

W. W. Norton & Company, Inc.
500 Fifth Avenue, New York, N.Y. 10110
www.wwnorton.com

W. W. Norton & Company Ltd.
Castle House, 75/76 Wells Street, London W1T 3QT

1 2 3 4 5 6 7 8 9 0

To my father

CONTENTS

ACKNOWLEDGEMENTS

This book owes a great deal to the work of many of my scientific colleagues. I could not have begun to write it without the sustained efforts of a small cohort of scientists who persisted long enough to discover the identifying codes written into the DNA of the Y-chromosome. They are, in no special order, Mike Hammer, Peter Underhill, Mark Jobling, Chris Tyler-Smith and Peter de Kniejf. Without their endeavours there would be no way of tracking this enigmatic chromosome and I am very grateful for all their hard work. I also want to thank Chris Tyler-Smith and Mark Jobling especially for letting me see manuscripts prior to publication. My own research team, in particular Eileen Hickey, Emilce Vega, Catherine Irven, Jayne Nicholson, Linda Ferguson and Lorraine Southam, have all helped to gather results which I have drawn on in writing *Adam's Curse*. Also from my research team, Helen Chandler and Shirley Henderson have pointed out new sources of material for the book which I might otherwise not have found. In the last year David Ashworth and the staff of Oxford Ancestors have helped me to finish off some lines of research and I am very grateful for that. Mark Crocker and Kathryn Churchley were very tolerant of my presence in their chromosome laboratory, and for that and their enthusiastic help I am grateful. I have also benefited, either through reading or in conversation, from the expertise and insights of Bobbi Low, Matt Ridley, Timothy Taylor, Robin Baker, Jeremy Cherfas, John Gribbin and Laurence Hurst.

Among the company of genealogists and other experts who have guided me through their perplexing and sometimes arcane discipline, Dr George Redmonds stands out as a fount of knowledge, wisdom and enthusiasm; and I would not have got anywhere on my researches into the Clan Donald without the help of Margaret Macdonald. Neither could I have done without the indulgence of the thousands of people who have allowed me to test their DNA and to reveal, sometimes, their intimate genetic secrets. In particular I want to thank Sir Richard Sykes and the chiefs of Clan Donald, who have all agreed to let me write about what I found out about their own Y-chromosomes. I could not have even begun the research without the help and permission of the Scottish Blood Transfusion Service. From the directors of the regional centres to the teams who accommodated us with patience and kind indulgence in donor sessions throughout Scotland, all were essential to the telling of this story. Along the way I have also benefited from the generous help of Mary Pontefract from Slaithwaite and the Lewis family from Poole. Among my friends, William James – now a fellow of even more Oxford Colleges than before – stands out as a wise mentor and enthusiast, always ready to entertain the wildest of scientific ideas (of which there are plenty in the pages to come). Lastly, I owe a very great deal to Janis Wilson and Sohani Hayhurst for helping me untangle the deeper mysteries of Adam's Curse.

I am lucky to have such excellent editors in Sally Gaminara and Simon Thorogood who, with patience, support and critical advice, moved *Adam's Curse* from a raw idea to a finished book. In a very practical sense Julie Sheppard has done the same with her astonishingly accurate transcriptions of my scribbled notes, for which I am very grateful. However, the final product owes everything to the eagle-eyed professionalism of Gillian Somerscales, my copy editor once again. And, of course, I owe a big thank you to my agents Luigi Bonomi and Amelia Cummins, who have kept my spirits up with an energy and enthusiasm which knows no bounds.

ADAM'S CURSE

PROLOGUE

A long time ago, when I was nothing more than a colourless clump of cells the size of a grape pip clinging to the dark inside of my mother's body, something happened to change the entire course of my life. Deep within my cells, a muffled detonation on one of my chromosomes triggered an unstoppable and irreversible chain reaction. A new genetic force pulsed through my minuscule body, throwing one cellular switch after another and resetting the coordinates of my embryonic voyage. Imperceptibly at first, degree by degree, I was diverted away from the normal course of development. Cells within my body laid aside one set of genetic instructions, unrolled another blueprint and set to work altering my small anatomy. Doors that had opened onto long corridors I was following were suddenly closed, and I could not turn back. Other doors opened that led me off in a different, unfamiliar direction, a direction which was eventually to set me apart from half of humanity. Seven and a half months later I was pushed out from my warm home into the blinding white light of the world. The very first words I ever heard defined what I had become. 'It's a boy.'

The corresponding announcement which greets every birth

1

colours the entire life of every one of us from the cradle to the grave. Sex is our principal badge, the first characteristic of any sort of personal description. The fact that we humans exist in two forms is so much part of everyday life, and always has been, that we rarely pause to question why this should be. Yet, the simple distinction between male and female divides our species into two perennially polarized camps separated on either side of a great canyon from whose rim we signal to each other and struggle to hear, but which we can never cross.

It is no secret that, underneath it all, men are basically genetically modified women. In this respect, our evolution can be regarded as a gigantic and long-running GM experiment. Its legacy has been to endow men and women with different and often conflicting sets of genetic interests, and to set off a powerful evolutionary spiral which has rapidly and sometimes dangerously accentuated the differences between the two sexes. This book is my explanation, as a geneticist, of the causes and effects of this endlessly fascinating yet often troublesome experiment which bewitches and entangles us all.

I have called the book *Adam's Curse* because the experiment which gave us men is not turning out too well just now, as any look at the newspapers easily confirms. Here are just two from the inside pages of today's editions.

POLICE HUNT VIOLENT LONER AFTER WOMEN ARE DISMEMBERED. A dangerous loner believed to have killed and dismembered two women was being hunted by police last night. Scotland Yard named Anthony John Hardy, an unemployed man in his mid-fifties, who lived close to where the remains of the women were found in Camden Town, north London. (*Daily Telegraph*)

MURDER CHARGE. Brian McCormack, 19, appeared before magistrates in Manchester charged with the murder of Jolyon Griffin, 28, who died on Christmas Day after being attacked on a city centre bus 11 days ago as he made his way home after a night out. (*The Times*)

In both cases, the suspect is a man. I would have had to search the papers for weeks to find a woman accused of a comparable crime. On the same day a far more disturbing yet not entirely unconnected story dominated the front pages:

BUSH SENDS 15,000 TROOPS TO GULF AS IRAQ ATTACK NEARS. America yesterday ordered its first full infantry division to the Gulf, prompting Pentagon sources to say that an attack against Iraq could be launched at any time. (*Daily Mail*)

It is a weary lament to lay most acts of violence and aggression, from the strictly local to the truly global, squarely at the feet of men. Yet the association is strong and undeniable. Women only rarely commit violent crimes, become tyrants or start wars. In *Adam's Curse* I explore the genetic explanation for this stark truth and point an accusing finger at the only piece of DNA which men possess and women do not: the Y-chromosome. There are other vital genes which, though both sexes carry them, are passed on only by women. These differences lie at the very heart of the genetic conflict between the sexes, set up by the great experiment, which resonates throughout our daily lives. Ironically, although the Y-chromosome has become synonymous with male aggression, it is intrinsically unstable. Adam is as much cursed as cursing. Far from being vigorous and robust, this ultimate genetic symbol of male machismo is decaying at such an alarming rate that, for humans at least, the GM experiment will soon be over. Like many species before us who have lost their males, we run the real risk of extinction.

The more I dug down, the more I realized that the two sexes are caught in a dangerous genetic whirlpool, playing out in the flesh irreconcilable conflicts embedded deep within our genomes. As much by luck as judgement, my own research on DNA placed me in a unique position to observe this primal struggle. I found myself with the means to follow the different genetic histories of men and women. I could listen to the messages carried by DNA and catch the whispers of old lives passed on by generation after generation of ancestors. When I finally woke up to what they were telling me, a lot of things that had made no sense at all started to fall into place. *Adam's Curse* is the result.

*

On a very practical note, sex and the reasons for it are fundamental to this book, and I use the word in several different contexts. Sometimes it refers to reproduction, sometimes to gender and sometimes to intercourse. I adopt this general usage to avoid, among other things, the angst of defining *exactly* what I mean by gender and to sidestep such literary absurdities as describing the shedding of pollen as any sort of intercourse. I hope the context will make my meaning clear.

1

THE ORIGINAL MR SYKES

As a geneticist, my professional interest in sex began over a decade ago when I first started to use that science to unravel some of the secrets of the human past. I chose as the instrument to navigate these mysteries a piece of DNA which is inherited purely down the female line, passed from mother to daughter for generation upon generation directly from our ancestors to the present day. This choice was made not out of any greater interest in women than in men on my part but because of its special properties. What this particular stretch of DNA revealed was not so much a history of our species as a history of women. And what a history it is. I was able to show that each of us is connected by unbroken maternal threads, traceable with DNA, to one of a few ancestral women living thousands, even tens of thousands of years ago. I was also able to track the movements of our ancestors across the globe and solve some of the riddles that had puzzled scholars for centuries – among them the origin of the Polynesian islanders, the fate of the Neanderthals and the nature of the first colonization of Europe by *Homo sapiens* before the last Ice Age.

I was well aware that, because the DNA I had used was maternally inherited, my interpretation of past events was based entirely on the genetic history of women and would need to be confirmed and complemented by an equivalent genetic history of men when that became technically feasible. However, I was confident that the main events had been interpreted correctly and that, although they might well be revised, the conclusions I had reached would not be substantially altered when the history of men came to be known. After all, men and women had to have been in the same place at the same time. I was quite content to leave unravelling the history of men to others and began to turn my attention to other projects. Then a chance event occurred that changed the course of my research and sent it spinning off in a new direction. And it brought the genetics of men right back into sharp focus.

As so often, the sequence of events began with a phone call – a call which was, in itself, nothing out of the ordinary. I work in the Institute of Molecular Medicine in Oxford as a professor of genetics, and from time to time I am asked to give talks on the subject to pharmaceutical companies. This particular call was an invitation from Glaxo-Wellcome (now part of Glaxo-SmithKline) to join a group of other scientists from Oxford at a conference at their company headquarters. Like many drug companies in the mid-1990s, Glaxo-Wellcome had realized that the discovery of new genes by the Human Genome Project, then well under way, would identify new targets around which to design drugs. If the genes for the big killers – heart disease, diabetes, cancer and so on – could be found somewhere in our DNA then they might show us what was going wrong when these diseases occurred, and new drugs could be designed to correct the mistakes. That, at least, was the theory.

What makes this particular invitation relevant to my story is that the chairman of Glaxo-Wellcome at the time was Sir Richard Sykes. As you can imagine, I was asked several times by the organizers from Glaxo-Wellcome in the run-up to the meeting whether Sir Richard and I were related. The only Richard Sykes I knew at the time was my own son; as far as I knew, their chairman and I were not connected at all. You can tell from Sir Richard's accent that he was brought up in Yorkshire, in the north of England. I, on the other hand, spent my childhood in London and have the accent to match. The only similarity between Sir Richard and myself, other than our both having trained as scientists, is that we have the same surname. I thought no more about it.

When I got into the car which had arrived to take me to the conference, the driver asked me the same question again. I don't know why, but this time, as I was about to repeat my simple denial, I suddenly had a thought. Maybe Sir Richard and I were related after all, *but without realizing it*. And, more to the point, maybe I could prove it by a genetic test. I asked the driver to wait, rushed back into the Institute, grabbed one of the small brushes which I used to collect DNA samples and ran back to the car. Sir Richard was going to be at the conference; I would ask him for a DNA sample and then compare it to my own. If he and I really were related then we would both share one very special piece of DNA. We would have the same Y-chromosome, that piece of DNA which every father gives to his son.

The following day, back in my laboratory, I took the small brush from its package. Invisibly attached to the nylon bristles were the cells that Sir Richard had brushed from his inner cheek the evening before. Though there were only a few hundred of them, they would be more than enough for me to get a genetic fingerprint of Sir Richard's Y-chromosome. Taking great care not to touch them, I cut the bristles away from the stem of the brush and dropped them into a small test tube. The cells had dried out overnight, but DNA is such a tough material that I had no doubt it would still be intact. After all, in previous research I had managed to get DNA out of human fossils over ten thousand years old, so I wasn't worried about a sample that had only been 'dead' for a few hours. Sir Richard's Y-chromosome lay at the centre of his cells and I needed to strip away the rest of the cell to get at it. Because DNA is so robust, I could use quite brutal chemistry to do this and the harsh treatment started straight away. I covered the cells with a few drops of water, then boiled them hard for ten minutes. This rehydrated the cells and burst through the delicate membrane that surrounds the nucleus, the very centre of the cell where his Y-chromosome was hiding. Now, after the boiling-water treatment, it was naked and exposed and could be minutely examined by the intricate molecular reactions that revealed its precise genetic fingerprint. I will say much more about this process later on, but for the moment all we need to know is that it worked perfectly on this important sample.

After a couple more days' work I had got Sir Richard's detailed genetic fingerprint from his purified DNA. Then, on my computer, I called up my own Y-chromosome fingerprint, which I had read several months previously. It resembled a bar-code, a series of dark and light bands that define a unique identity. I lined it up with Sir Richard's and went along the pattern, one bar at a time. They were all exactly the same. Our Y-chromosomes matched perfectly.

It was proof that the two of us were related. But how? Both Sir Richard and I had inherited our Y-chromosomes from our fathers, who had inherited it from theirs, who had inherited it from theirs, and so on back in time. Our Y-chromosomes were tracing two direct lines of paternal ancestry which went further and further back into the past. Since our Y-chromosomes were identical, this had to mean that the lines we each traced back through our fathers, our grandfathers, our great-grandfathers and so on converged at some point on just one man. This man, whoever he was, was our common paternal ancestor, a man to whom both Sir Richard and I could trace, through our Y-chromosomes, an unbroken genetic link. Since we had also inherited our surnames via the same route it was extremely likely that this man, our common ancestor, was also called Sykes. At a stroke, our Y-chromosomes had proved a connection between us that no documents had ever suggested.

Even now, years later, we still don't know precisely how we are related, and it might take years of patient work to trace the connection through records of births, marriages and deaths – if it could be done at all. But somehow that doesn't seem to matter. The genetic thread is direct and continuous, regardless of the generations of men through whom it has passed.

Having shown the genetic link between Sir Richard and myself, I began to wonder how many other people called Sykes were similarly related. Could it possibly be that we all were? I am ashamed to say that at the time I knew next to nothing about the origin of my surname. About all I did know was that my grandfather had been a soldier in the First World War and that his family had come from somewhere in Hampshire in southern England. As far as I was aware, there was no connection with Yorkshire that could possibly link my family to Sir Richard's. Had my family moved from Yorkshire to Hampshire at some time in the past? Or had Sir Richard's gone in the opposite direction, from Hampshire to Yorkshire? Where did most of the Sykeses live anyway? I didn't have a clue.

About this time I got a letter through the post at home. This was an invitation to purchase the grandiloquently entitled *Book of Sykes*. Normally this kind of circular would have headed straight for the bin but, curious for the first time to know more about the name, I sent off for the book. Expecting an in-depth exploration of the history of the family, I received instead a folder with some very general blurb on surnames, a suspicious-looking coat-of-arms and, at the back, a list of names and addresses of Sykes men, arranged by county. Had I been interested only in the name, I would have been disappointed. But, though I was none the wiser about its history or origins, the list at the back was just what I needed. Looking through it I saw at once that there were far more Sykeses living in Yorkshire than anywhere else. So it looked as though it had been my ancestors who had been the ones to move rather than Sir Richard's. I picked 250 Mr Sykeses at random from

Yorkshire and the neighbouring counties of Lancashire and Cheshire, and wrote to each of them asking for a sample of his DNA. Since I was a Mr Sykes writing to other Mr Sykeses it did not feel so much of an intrusion as it would otherwise have done. Enclosed with each letter was a DNA brush, and within a month I had received back about sixty samples of Sykes DNA.

Let me say at this point that I now know from bitter experience that, although there is nothing more fascinating than your own family history, there is nothing more tedious than someone else's. So please forgive me while I tell you some things about the Sykes family. I do it only to illustrate, not to inform, and when I have finished you are free to forget all about us.

I had done a little more research on the name and discovered that Sykes derives from the Yorkshire word 'sike', which is a particular kind of moorland stream. No magnificently gushing torrent, this; a sike is more of a slow trickle in a ditch, and sikes often marked boundaries between adjoining plots of land. If I was hoping to prove that all living Mr Sykeses were related and ultimately traced their origins back to a single founder, this news was not encouraging.

Other than among the aristocracy, most English surnames were introduced around the thirteenth century, principally as a tool of estate management. By this time pretty much the whole country was divided into large feudal estates, a direct legacy of the Norman invasion in 1066 by William the Conqueror, who handed them out to his friends and supporters. A feudal lord controlled all the land on the estate and distributed the agricultural land among the tenant farmers, whose rents kept him and his immediate family in the grand style to which they very soon became accustomed. This was a highly regulated structure, and detailed records were kept – of which many still exist – listing the size and rental of each parcel of land along with the name of the tenant.

The trouble was that, without surnames, it was almost impossible for the estate officers to keep track of events. Within a small village, where everybody knew everybody else, it was easy for the residents to cope with several people having the same name. They knew them as individuals and often by a nickname as well. But the estate managers had huge difficulties. It was often impossible to tell which John or Adam or Mary or Maud was which. Their solution was to differentiate between people with the same name by adding another name – a surname. Soon afterwards these new surnames became hereditary. By the middle of the thirteenth century, tenant farmers were permitted to pass on tenancies to their sons when they died, so it was natural under the circumstances for the surname to become hereditary, just like the tenancy itself. It was this very practical aspect of medieval book-keeping that lay at the origin of most English surnames. From these bureaucratic beginnings, eventually every man was given a surname; on marriage, women took the names of their husbands. Sometimes these surnames were derived from an occupation – like Carpenter, Smith or Butcher; sometimes they evolved from a nickname, often a descriptive one, such as Redhead or Smallpiece. Other surnames merely added '-son' to the name of the father to form patronymics like Johnson or Adamson. A fourth category of names were derived from a feature of the landscape – Hill, Bush, Wood and, in Yorkshire, Sykes.

That was the discouraging prospect. Since there were literally thousands of sikes in Yorkshire, the chances that only one man had decided to adopt 'sike' as his surname seemed extremely slim. Even though the Y-chromosome results certainly suggested that at least Sir Richard and I were descended from the same man, the likelihood of a large proportion of the random samples I had collected from other Mr Sykeses being similarly related seemed remote indeed. However, when I deciphered their Y-chromosome fingerprints the results were truly amazing. Fully half of the Sykes samples, randomly collected from the three counties of Yorkshire, Lancashire and Cheshire, had exactly the same fingerprint. There was only one possible explanation for these spectacular and unexpected results. The volunteers, including Sir Richard and myself, who have the same Y-chromosome fingerprint, must have inherited it from a common ancestor. All of us must be able to trace a direct father–son lineage back to one man. But who was this man? Was he the original Mr Sykes? And, equally important, what about the other half of the sample, the men who did not share this Y-chromosome fingerprint?

Let's tackle the second question first. The Y-chromosomes that didn't match the 'Sykes' fingerprint, as I was now starting to call it, were split into two categories. A few, while not exact matches, were very close to the 'Sykes' Y-chromosome. The others had very different 'bar-codes' and were completely unrelated to it as far as I could see. Not only that, they were not obviously related to one another either. There were no other clusters of related Y-chromosomes to suggest we had found the descendants of a second 'original' Mr Sykes. What explanation was there for this pattern, where half of the Sykes men shared the same Y-chromosome fingerprint and the other half had a mixture of Y-chromosomes with no obvious relationship to each other?

At this point in the narrative we need to introduce the factor to which geneticists politely refer as 'non-paternity' – the term used when a child's father, the name on the birth certificate, is not the biological father. When a son bears the surname of his father but does not carry his genes there are only a few explanations available. The most straightforward, and innocent, is that the son has been adopted and taken the surname of his adoptive father. Of course, the same happens to adopted girls, but they will most likely not transmit this name to their children and they will certainly not pass on a Y-chromosome either. Y-chromosomes are only ever passed between father and son. Women just don't have them. The second explanation is that the entire family adopts a new surname. This was not a common practice in medieval England but it certainly was in Scotland, where a man often took the name of the clan chief on whose lands he lived or in whose army he fought without being related to him. That leaves us with the third and final explanation for the discordance between surname and Y-chromosome – infidelity by, or possibly rape of, the woman. Biologists have a rather more brutal name for it – extra-pair copulation. If a woman has a child with a man other than her husband and if that child is brought up within the family and is given the family name, the link between name and genes is broken. If the child is a boy, he will inherit his father's surname but not his Y-chromosome. That will have come from his mother's lover, or from her assailant, and not from her husband. When he has sons of his own, it will be this man's Y-chromosome that is passed on. Even if there are no non-paternity events in later generations, the link between the Y-chromosome and the original surname cannot be rescued. It is severed for good.

From our admittedly limited survey, the Sykes Y-chromosomes fall into two roughly equal categories. The first group were very closely related to one another and are almost certainly inherited, without interruption, from one original Mr Sykes. The other half of our volunteers had inherited Y-chromosomes which are very different from the 'original' Sykes chromosome and also from one another. These Y-chromosomes could have become associated with the name through infidelity, rape or adoption at some point since the name started. Or they might be the Y-chromosomes of several different 'original' Mr Sykeses, each passed down to the present day through a direct paternal line unbroken by non-paternity events. From this evidence alone it is impossible to tell the difference; however, if they were from different originals, none of them had done anywhere near as well as the main Sykes chromosome.

Though there is no way of formally distinguishing these different possibilities, I thought we should be able to work out a figure for, if you like, the accumulated rate of non-paternity events. This would be an estimate of the proportion of non-paternity, of whatever type, that had occurred since the thirteenth century to give us the present pattern, with half the Sykes men sharing the same

Y-chromosome signature and the other half showing a mixture of apparently unrelated genetic fingerprints. I needn't trouble you with the calculation; the answer comes to 1.3 per cent non-paternity events per generation. It means that for over seven hundred years, the average rate of adoption and illegitimacy could be only just over 1 per cent in each generation. Had it been much higher than this, the pattern we see among modern Sykes Y-chromosomes would have disintegrated long ago. Put another way, it means that 99 per cent of Mrs Sykeses have been very well behaved, or very lucky, for the last seven hundred years. In fact, since this figure also incorporates the possibility of other independent founders of the name it is the *maximum* estimate for non-paternity, and when you bear in mind that some of these events would have been genuine adoptions, the illegitimacy rate falls yet lower still. How does that compare to rates of non-paternity these days? Surprisingly, there is no universally accepted value for the current rate, but the range of estimates (5–30 per cent) in different studies in the UK are all much higher than the historical values obtained from the Sykes results.

Even with the difficulty of distinguishing the influence of non-paternity events from that of different independent founders, the overall result was staggering. Most, if not all, the volunteers from the three counties of Yorkshire, Lancashire and Cheshire had got the name from one man. And half of them still carried his Y-chromosome. Had I been incredibly lucky with the name Sykes? I don't think so. Over the past two years I have replicated this study with dozens of names. Not all of them show as tight an association between surname and Y-chromosome as Sykes, but most do and some are even more impressive. In one name, to which I shall return in a later chapter, fully 87 per cent of present-day holders have the same or very closely related Y-chromosome. From what I can see so far, the majority of surnames, in England anyway, are very clearly linked to one or a very few Y-chromosomes.

Of course, there had been luck involved: not so much in that no other name would have worked as well – it would – but for a completely unscientific reason. Had the chairman of Glaxo-Wellcome not been a Sykes, I would never have thought of doing the study in the first place. A second piece of good fortune was that Sykes is a Yorkshire name and Yorkshire just happens to be the home of one of the best surname experts in the whole of England – Dr George Redmonds. Without George, the Sykes Y-chromosome study would have ended up as a cold and formal scientific report; interesting, to be sure, but with no real connection to the history and the landscape that I was now aware had been the home of my genetic ancestors for the best part of a thousand years. Sykes country, as I now felt entitled to call that part of West Yorkshire south-west of Huddersfield, is a landscape of barren moorland intersected by steep-sided river valleys. From the top of the high moors the area looks almost deserted, with hills rolling away into the distance in every direction. Slightly lower down the slopes are the hamlets, the clutches of weavers' cottages each clustered around a farmhouse. Lower still, confined to the valley floors and completely out of sight from the hilltops, are the old mill towns, fully urbanized, noisy and dirty.

George lives high up on the moors and his knowledge of the area – its landscape, its history and particularly the histories of its families – is nothing short of encyclopaedic. A drive round this rugged landscape with him brought it vividly to life. The unnoticed line of a broken dry stone wall on a distant hillside became the failed attempt of a medieval farmer, pushed higher and higher out of the valleys, to cultivate the poorest land. One craggy peak – Wolfstone Heights – is no longer just a name on the map but recalls a time, not so very long ago, when there really were wolves living on the moors.

George and I first met when we were making a series of radio programmes for the BBC on the subject of surnames, genes and genealogy (produced by another Sykes – Sandra) and George began to search for the earliest records which mention the Sykes name.

Within a short space of time, he had unearthed a reference in the court rolls of 1286 to a Henri del Sike. George showed me some of these records, and their condition is quite remarkable. Inscribed on parchment, made from calf skin, they are strong enough, even after several hundred years, to be handled without disintegrating. Had they been written on paper instead, they would have crumbled into dust long ago. The particular court record that George had found referred to a tenancy dispute involving Henri del Sike in lands near to the village of Flockton, a few miles south of Huddersfield. The village is still there, and there are still Sykeses in Flockton, but a quick trawl through the electoral roll showed that there were far more in the small town of Slaithwaite, about nine miles distant. Slaithwaite, George already knew, was a much younger settlement than Flockton. It is situated at the bottom of a steep-sided valley, on the banks of the River Colne. These valleys were thickly wooded in medieval times, marshy and full of wild animals. This made them difficult to farm, so the hamlets and villages were established higher up on the valley sides where the land was well drained and largely clear of trees. It was only much later, in the eighteenth and nineteenth centuries, when weaving and textile manufacture became industrialized, that the valley floors became densely settled. The dark mills, thirsty for water to wash the wool and to feed the steam engines that powered the looms, needed to be built close to rivers.

The obvious question which George asked was this. Were the earlier Flockton Sykeses related to the Sykeses of Slaithwaite? He had discovered evidence of Sykeses living between the two settlements in the fourteenth century and had found a convincing explanation of why they might have moved away from Flockton. The Black Death – bubonic plague – had scythed its way through the population of Europe first in 1348 and then in subsequent epidemics of diminishing ferocity over the next hundred years. The initial epidemic killed between one third and one half of the population in the space of eighteen months. It is hard to imagine the terrible effects of an epidemic on that scale among our ancestors. No family escaped as fear and death swept across the land like a swift black shadow. After the epidemic burned itself out, unable to find sufficient susceptible victims still living to sustain it, the survivors found themselves in a new economic landscape. Faced with an acute shortage of labour, the feudal lords were forced to improve the wages and conditions of their tenants and serfs. Land cleared of its occupants by the Black Death became available to new occupants. In George Redmonds' opinion, it was the opportunity to settle on new land which had persuaded some of the Sykeses to leave Flockton and seek their fortunes elsewhere. Now the genetics suddenly gave George the opportunity to test out his idea. If the Sykeses in Slaithwaite had come originally from Flockton, then the Y-chromosomes of the two groups ought to match. I wrote to the Sykes men in Slaithwaite and Flockton asking for samples of their DNA – and, when we analysed their Y-chromosomes, we found that the genetic fingerprints were absolutely identical. George's hunch had been right, proved beyond any doubt by this new genetic test.

I wanted to see the original site near Flockton which George had linked to the very first Sykes in the records. It was a cold day in early April when we got out of the car next to a stream which ran along a valley floor. The trees were not yet in leaf and great oaks stood naked in the green fields opposite. These pastures led up to a ridge about three hundred yards away where the village of Flockton itself is strung out along the brow of the hill just as it has always been. To our left, beyond a dry stone wall, an uncultivated croft was alive with the golden flowers of kingcups in the boggy ground close to the water. The stream itself was clear and bubbling but the bed of the river was dead, choked by the rust-coloured deposits of ochre, the still polluting effluent of long-abandoned iron-ore mines.

A track led off across the stream and George led me down it between tall poplar and aspen trees which hugged the water. At a bend in the watercourse stood the ruins of an old mill, abandoned long since. George had pinpointed this particular spot by finding that Henri del Sike had the tenancy of land on both sides of the stream, which lay in different parishes. There was no sign of the farmhouse which my ancestor, the very first Sykes, had occupied, but even so, it felt quite extraordinary to be here. Looking around at the old mill, the track and the stream, it seemed that nothing in the landscape had greatly changed. Nor had it. The field and croft boundaries were as they had been in the late thirteenth century when Henri del Sike was living here. As I stood, I could almost hear the voices of children – my ancestors – laughing as they threw pebbles into the stream. Without the DNA evidence, it would have been an interesting enough experience to see where the first recorded Mr Sykes lived. But I would have felt detached from it. I would have known there was a connection of a sort between the place and me, but it would have been a connection made through the mind, the rational conclusion of a process that matched the name on my birth certificate with another name on a piece of yellowing parchment. But to know that the Y-chromosome that I carry in all my cells had actually been here, in this place, in the fields beside the stream, was a completely different sensation. Now it felt as if I were experiencing the history of a real part of myself, a place where some of me had actually lived. And, of course, it had.

2

THE LONELY CHROMOSOME

Although I had established a link with my ancestor who had lived by the side of a Yorkshire stream over seven hundred years ago, I had never actually *seen* the chromosome that had travelled from him to me down through the generations. I knew it only as a bar-code, a series of lines on a computer screen. To be sure, the detail of that bar-code had led me to connect myself with Sir Richard and all the other Sykeses from Yorkshire, but I still felt it was strangely anonymous. This is a piece of DNA that had come down to me by a very special route. It had been given to me by my own father, who had received it from his father, who inherited it from his father. We receive DNA from all of our ancestors, but the Y-chromosome traces such an important history that a simple bar-code, visually similar to what you might find on the side of a packet of frozen peas, doesn't do justice to its very special nature. This was the DNA that had made me a man. And not just me – every man owes his maleness to his Y-chromosome. If I was going to explore its deeper secrets I wanted to know what it looked like, to see it with my own eyes. The central character of such a powerful drama must have a face.

Though it was many years since I last made a chromosome

preparation, my friends at the Clinical Genetics Laboratory in Oxford must have thought me sufficiently technically competent to let me try. Their lab is in the Churchill Hospital in Oxford, a short cycle ride from my own laboratory, but with a very different architecture from the modern building in which I work, still struggling to escape from its prewar design of endless corridors. Immediately I got past the entrance, I was back to my early childhood, lying on a trolley, my arms folded over my chest, being wheeled into the children's isolation ward. Though this was not the same hospital, the long, brightly lit corridors and the faintly sweet smell of boiled vegetables which hung in the air returned me at once to that day long ago when I was admitted. I was nine years old and I had contracted bacterial meningitis. I was in hospital for three weeks, and only years later did I realize how lucky I had been to survive. I remember vividly that my greatest fear was not of a relapse or any lingering long-term effects from the infection; it was of the blood test which the nurse told me to expect before I was allowed home. It was only going to be a finger-prick with a lancet, a tiny drop of blood, but the prospect was quite enough to terrify me. In the end I escaped without ever having the test. During my work on DNA, I have taken finger-prick blood samples from thousands of other people, including children, and I always remember how scared I was myself.

Today, however, I am going to have a proper full-size blood sample taken from the vein in my left arm. Luckily, that vein is large and conspicuous, a blue-grey tube that nobody ever misses. But, even now, I still feel slightly sick as the needle slides in and the dark red blood flows into the vacuum tube. I take the tube, now full of my blood, along the corridors to the genetics laboratory. Within the tube a billion red blood cells hang in suspension. These tiny red globules, whose job it is to keep my tissues supplied with oxygen, contain no chromosomes and play no further part in the search for my own Y-chromosome. Among the red cells, outnumbered by a thousand to one, float my white blood cells. Their job is to protect me from infection; to recognize a virus or a bacterium as foreign, an unwelcome intruder that must be eliminated. Once the white blood cells have identified their target, they draw on a fearsome army of weapons to destroy it. Among other things, the white cells begin to manufacture antibodies, proteins that are exquisitely designed to envelop specific strains of bacteria or viruses that have broken through into the bloodstream. Other white cells then engulf the antibody-coated invaders, chew them up and spit them out. To produce antibodies and to destroy invading micro-organisms, the white cells need to have genetic instructions which, as we shall see, reside on the chromosomes. So, unlike the red cells, which jettisoned theirs as soon as they had read the haemoglobin instructions, the white cells have held onto their chromosomes. However, they are invisible except for a few hours surrounding the moment of cell division when they appear briefly before once again melting into the background of the cell. So, if I am to stand any chance of seeing my own Y-chromosome I need to persuade my white cells to begin dividing. The first thing I must do in the lab is to take the tube of blood into the culture room. I put on a white coat, slip on a pair of surgeon's gloves and enter the room. To one side stand two illuminated cabinets. The room hums with the deep roar of filtered air being pumped into the cabinets and then blown outwards to protect the blood cultures from infection. One of the responsibilities of the Clinical Genetics Laboratory is to screen the cells of sick children, and through this room pass the hopes and fears of parents as these cells are analysed for defects in their chromosomes that might explain the children's mystifying symptoms. Through this room, too, pass the cells taken from the placental sac during amniocentesis that are coaxed into division to reveal the extra chromosome that foretells a young life burdened by Down's syndrome.

The first stage of my Y-chromosome preparation is brief and functional. I make up a few millilitres of culture medium which contains all the nutrients my white cells will need to survive and, I hope, begin to divide. There is glucose mixed with the cocktail of various metals, in minute concentration, that the cells require to fuel their metabolism. There is sodium bicarbonate to keep the cells at a precisely neutral pH, the right balance of acid and alkali, and a coloured dye to warn me if this balance is upset. If the colour is a gentle orange then the balance is fine. If it changes to a vivid pink or a bitter yellow, rapid action is required to restore the pH balance and rescue the cells. The medium now prepared, I augment it with a mixture of different antibiotics to ward off infections from the air, a little heparin to stop it clotting – and a magical ingredient, an extract of the phaseolus bean, which sends the white cells off into a frenzy of division. A chemical in the bean extract reacts with molecules on the surface of the white cells in a way that mimics the effect of a massive bacterial invasion, prompting the white cells immediately to begin to divide ready to mount a counter-attack.

Thus far I have not worked with my own cells, only prepared the medium in which they will grow. It is a modern precaution against the very remote possibility that my white cells might become inadvertently infected with a virus from a culture of someone else's blood growing in the laboratory. If that happened and if, by accident, my own cells, now infected, got back into my own body through a cut or a needle injury, because they are mine they would not be recognized and rejected by my immune system, and I might become seriously ill; so I am not allowed near my own cells while they are still alive. They are in the care of Kathryn Churchley, on whose dextrous hands my growing cells now depend. In the unlikely event that she accidentally injects herself with my cells, they would be rapidly destroyed by her own immune system and she would come to no harm. She adds a few drops of my blood to a small tube of the culture fluid and gently rocks it to and fro a few times to mix the ingredients. Then she quickly opens the door of an incubator, like a small oven, and puts the tube inside. It will be kept here at body temperature for the next three days while the white cells divide. There is nothing more to be done now. Either the cells will grow or they will die.

After three long days, I arrive back in the lab knowing that, if all goes to plan, today is the day I will see my chromosomes for the first time. Kathryn takes the tube of cells from the incubator. The white cells are still invisible, the colour of the liquid still dominated by the haemoglobins of my red blood cells. Though they vastly outnumber their pale companions, they will not have divided over the past three days. Having abandoned their chromosomes, they cannot grow. A few hours before my arrival, Kathryn had added a drop of the drug colchicine to the cell culture. This ingredient, distilled from the underground stems of the autumn crocus and used as an ancient treatment for gout, destroys the delicate filaments that drag the chromosomes apart at the very last step of cell division. While the colchicine is there, the chromosomes are frozen at this final stage. For the few hours they have been exposed to the drug, my white cells have been trying to divide, only to be halted at the very last moment. As more and more cells reach this point, the number of cells arrested at the final frontier of their life cycle gradually builds up until there are thousands of them. Their chromosomes are all frozen at the same moment. And that is just how we want them: at that phase of their extraordinary lives when they are sufficiently compacted and condensed that we can actually see them.

Before I can get my first glimpse, there is still a lot to do. To harvest the cells, and separate them from the culture fluid that has nourished them for the past three days, Kathryn puts the tube into a small centrifuge. The machine growls into action and starts to turn. As the speed increases and the tube spins round, the cells begin to be driven by the rapidly increasing centrifugal force down through the fluid to the bottom of the tube. By the time the centrifuge reaches its maximum speed, the tube is spinning at twenty revolutions a second and the cells, red and white alike, are hurtling towards its base. After five minutes they have all collected at the bottom, the centrifuge sighs as the motor slows and the tube comes to a rest.

There, tightly packed in a dark red pellet at the bottom of the tube, are my blood cells. They are still alive, but not for very much longer. Kathryn draws off the old culture fluid with a pipette. We no longer need it. In its place, she adds instead a clear salt solution. The salt concentration is finely judged. It is a little less concentrated than the corresponding solution within each of my cells – but not by much. As soon as Kathryn mixes my cells with the solution they start to swell as they suck in water by osmosis. We cannot see this, but as more and more water floods into my cells, the membranes that surround them tighten and stretch like over-inflated balloons. The red cells have the weakest membranes and they begin to burst, spilling their cargo of haemoglobin into the clear solution. The white cells have a slightly tougher membrane, a slightly thicker skin. That is the reason why the salt solution has to be formulated so precisely. If it were only slightly more dilute, the white cells would be forced to swell even further and they too would burst. If it were just a little more concentrated, the red cells would be able to withstand the swelling pressure. Only at exactly the right concentration, at precisely 4.19 grams of salt per litre, no more and no less, do the red cells burst open and the white cells remain intact for us to collect.

Once more the tube is carefully placed in the centrifuge and spun round and round. The centrifuge sighs and slows to a stop, and Kathryn takes the tube out and holds it up to the light. I can see a small dark red pellet at the bottom of the tube, but it is much smaller than before. These are the red cells that had not burst. Lying on top of this dark red plug I can see a thin, faint, greyish-white layer. These are my white cells, still outnumbered by red cells that had refused to puncture, but there nonetheless. There is no more we can do to separate the white cells from the red while they are alive. Now they must all die.

Kathryn pours off the red liquid, the haemoglobin and fragments of red cell membranes too light to be spun out by the centrifuge. The small red pellet remains. She shakes it, to suspend the cells in the small amount of liquid, then fills her pipette with the agent that will end their life. She calls it 'fix' – short for fixative. It is a mixture of alcohol and acid, a lethal combination that will extinguish all signs of life in the cells that remain. Deftly swirling the contents of the tube in her left hand, she adds one drop of 'fix' with the right. As the drop slides down the inside of the test tube and mixes into the swirling suspension, a strange transformation comes over the contents. The cells that had helped sustain me until they left my body a few days ago are dying. The colour of the solution changes, at the instant of their death, from a vibrant, vital red to a sickly olive-grey as the iron atoms in the haemoglobin shift their molecular position. The colour of life has changed to the colour of death.

Kathryn has guided my cells through all these perilous procedures but, now they are dead, I can take them on from here without any fear of inadvertent infection. I add a few more drops of 'fix' and, grasping the precious tube, leave the culture room with its breathing cabinets, its spinning centrifuges and its bank of warming incubators and move back to the main laboratory. This is the place where the cells will give up their chromosomes. Suspended in a few drops of liquid, my own white cells are still intact. Holding the tube up to the light, I see the faintest of white smudges on the bottom of the tube. It looks so small, so insignificant: a tiny dusting of something that it is hard to believe holds the genetic secrets of my ancestors.

I must now explode my own white cells and spread the chromosomes within them across the surface of a microscope slide. This is the step, the most difficult to judge, at which art and science most closely touch. The cells must be dropped from a height of a few inches onto the glass slide. Weakened and filled to bursting by the earlier treatments, the cells rupture on impact. The chromosomes are thrown out onto the glass and stick there. The art, the skill, comes in dropping the cells from just the right height, with just enough force that the chromosomes are scattered – but not too far.

Too brutal a treatment now will fling the chromosomes all over the slide; too gentle a touch and the chromosomes will be tightly bunched, lying in a tangled heap. I take the glass slide and gently blow across it. A thin layer of condensation settles briefly on its surface. Immediately I let one drop fall from the pipette. It spreads across the surface of the glass slide. I add one drop of 'fix' and it moves instantly across the glass, propelled by a combination of surface tension and humidity, spreading the chromosomes as it goes. I dab off the excess liquid with a tissue and wait for the slide to dry. After about a minute I can see a faint ring of grey surrounding the point where the first drop hit the slide. These are my cells and among them, I hope, are the scattered chromosomes.

Microscopes are the way into a different world, a world of bizarre creatures and fantastic shapes that are all around us but invisible to our eyes. My grandfather was an inventor and amateur scientist, and he gave me an old brass microscope in a mahogany box when I was a boy. Through the lens of this ancient instrument I had seen the alien shapes of pollen, the interlocking scales of a butterfly's wing and the mysterious green globes of minute pond algae, each one as strange as any device of the human imagination. My old brass microscope was beautiful, but it was not, in truth, particularly good. I could never see, for example, any detail in the cells of my own blood. They were shimmering points of light but nothing more than that, even at the highest magnification. The microscope beside me now in the genetics laboratory is less beautiful but has an incomparably better optical performance. I slide the glass under the lens and look through the binocular eyepiece, adjusting its width to fit the separation of my own eyes. Slowly turning the knurled wheel on my right, I focus down on the slide. The blurred images of spheres become sharp and, as they come into focus, I find myself looking at about a hundred trans-lucent circles against a greenish background. These are my intact white cells, the ones which had not ruptured. There are a dozen small clumps of dark flecks among the intact cells. These flecks, barely visible at this magnification, are my chromosomes, the bare embodiment of my genetic identity – and I am seeing them for the very first time.

I look again and change the lens to a higher magnification. Now I see only a few clumps of chromosomes. How very small these things look. I am more used to describing the vastness of the genome, accustomed to marvelling at the three thousand million DNA letters from which it is built. I know very well the enormity of the technical achievement of the Human Genome Project in deciphering the whole sequence, the millions upon millions of smaller sequences that had to be overlain and stitched together to give the final unimaginably long sequence. I was used to describing the genome in metaphors of vast distances – pointing out, say, that if all our DNA stretched from London to San Francisco, a typical gene would only be one inch long. But here, as I peer through the microscope, what strikes me is how very small my chromosomes really are; and how unimaginably compressed the two metres of DNA which each cell contains must be to be crammed into this tiny genetic bouquet.

Although I can see them clearly, there is not enough detail for me to identify individual chromosomes or to pick out my Y-chromosome from the others. For that, I need to take them through more arcane procedures, none completely understood or capable of totally rational explanation, which will stamp each chromosome with an identity. The recipes, every bit as mysterious as the secrets of any ancient craft, have been handed down from master to apprentice for decades. First I take the slide and place it in a hot oven to be 'cured' as if it were a side of bacon. After resting there overnight, the now cured chromosomes are ready to be dyed to reveal their individual identities, and I am in early the next day to set up three square glass dishes on the laboratory bench. I fill the first with an orange liquid. Despite its toxic appearance, this is a gentle fluid which prepares the chromosomes for the later, more aggressive steps of the process. I put the glass slide, with its delicate layer of chromosomes clinging to its surface, into a stainless steel rack and plunge it into the orange solution. I set a timer for exactly three minutes and fill a nearby sink with warm water.

The second glass dish contains the vital ingredient responsible for the dark and light pattern of bands that we will use to tell the chromosomes apart. This is trypsin, the enzyme made by the pancreas to digest proteins on their way through the small intestine. Nobody really knows precisely what it does or exactly how it works. It probably scours away some of the protein scaffold around which the DNA is wound, exposing parts of the chromosome to the stain while protecting others. In this respect it is rather like batik, where beeswax shields the parts of the cloth that are not to be dyed. But exactly how the trypsin treatment creates the chromosome pattern is a mystery. Certainly it is a matter of judgement, like so many of the other steps. Too short a time exposed to trypsin and the chromosomes will stain uniformly, without showing any bands of light and dark. Too long an exposure to the enzyme and they will fall apart as the trypsin digests the protein scaffold that holds them together. Kathryn always uses twenty-three seconds. Not twenty-two or twenty-four but twenty-three seconds. So that is what I do. The slide comes out of its orange bath; I rinse it in the warm water in the sink, then plunge it straight into the trypsin. Exactly twenty-three seconds later, I take the slide out of the trypsin bath and immerse it in the warm water once more.

The third and final glass dish contains the stain itself. The colour of blue-black ink, it is a strong solution of a chemical dye called Giemsa, after its discoverer Gustav Giemsa. I place the slide carefully into this inky liquid, deep and impenetrable. Beneath the surface, the dense stain is attaching to the sections of chromosome etched free of protein by the trypsin. These sections will be the dark bands. Where the trypsin did not have time to dissolve the protein scaffold, the stain cannot penetrate. After its three minutes in Giemsa stain, I lift the slide out and plunge it into the water for the third and last time. The dark stain fills the sink, so concentrated is it. I dry the slide, protect it with a wafer of the thinnest glass imaginable and take it across to the microscope. Scanning the field, I pick what looks like a good cluster of chromosomes. From what I can see they are nicely spread out. With this cluster at the centre of the field, I click a high-power lens into position and look through the eyepiece. There they are – my chromosomes. This time they are each marked with a striated pattern of light and dark bands that cut across their length. These are the bands that reveal their individual identities. Now the volumes of my genome have titles on their spines. The longest chromosomes lie across the field, bent at their centres like battleships broken at Pearl Harbor. The smaller chromosomes point stubby fingers towards each other, each one now badged and identifiable. The smallest, barely visible through lack of stain, seem lost and incidental.

I scan the chromosomes by eye and start to pair them by appearance. First by size, long or short; then by their pattern of light and dark bands. Each one, with two important exceptions, has a twin somewhere else in the field of view, and I mentally cross them out as I find them. I am looking for the chromosomes that have no twins, the chromosomes that have come from only one parent, not from both. As I go through this process of elimination, two chromosomes begin to stand out, dissimilar in size and band pattern from any other on the slide. The smaller of these sits at the edge of the field, slightly adrift from the others. It has no partner, nothing with which it can be matched.

This is my Y-chromosome, the bearer of my maleness and the token passed unaltered down from a long line of fathers. This is the chromosome I have come to see. I see it in my own father, as he leads his RAF squadron in the Second World War. I see it in my grandfather, fighting in the trenches and wounded in the battle of the Somme a generation earlier. Before that I don't know where it was – except that seven hundred years ago it was in Yorkshire, beside the brook at Flockton. Back beyond that it vanishes into the mist.

My other chromosomes, lying contentedly on the glass slide, have come down to me from a mixture of ancestors. Theirs is a cacophony of different sounds, both male and female, individual voices are drowned in the throng of noise. It is only my Y-chromosome that now speaks with a single voice, one that has come to me from generations of men. It stands alone, a perfect copy of the chromosome that lived in my father and in my father's father and in a thousand others of my paternal ancestors stretching back to thirteenth-century Yorkshire and way beyond, back through thousands upon thousands of men into the far distant past. I stare at it, imagining its long journey from distant ancestors, alone and set apart from all other chromosomes. What is it that makes the Y-chromosome so unusual and also so very special?

3

RIBBONS OF LIFE

I have drawn you into this story with only the briefest of intro-
ductions to its main characters, the chromosomes and their vital
cargo – the DNA that makes us what we are. DNA, the lexicon of
heredity, is a code; and it is among the very simplest codes imaginable, written in just four letters. It is the precise sequence of these
letters which matters. But what is sending and what is receiving
this coded message? The receivers are the production units within
each cell, which fashion proteins from amino-acids. On receiving
its instructions from DNA, this machinery automatically begins to
make new proteins according to the coded message. The proteins,
in their turn, weave and build the human body and then supply the
intricate web of enzymes and hormones to keep it going. Though
this process is unbelievably complicated in the detail, the basic
principle adhered to by the interpreters of genetic instruction, the
cells, is very straightforward: read the instructions and do what
you are told. The cells have no powers of veto or amendment. They
only obey, even if the message spells out their own death. They cannot rebel. In utter contrast, the answer to the question of what is
sending the coded signals is in detail simple and straightforward.
Yet beneath the basic mechanics the answer is deep, mysterious and

awesome. Before we embark on the exploration of these deeper mysteries, whose influence affects us all in ways we barely comprehend, let us begin from what we know by observation.

Our DNA is arranged as a series of immensely long molecules, each of which is a physical embodiment of the code itself. In early 2001, the almost complete sequence of human DNA was deciphered and published – a truly astonishing technical achievement. This showed the precise order in which the four DNA letters (A, G, C and T) which hold the code appear and reappear. It is immensely long – about three thousand million letters in all – and commentators struggled for the equivalent numbers of volumes of *Encyclopaedia Britannica* it would take to match that amount of information. However, most of our DNA is not doing anything useful, and the thirty thousand or so genes – the real meat of the genetic instructions – are outnumbered by vast stretches of so-called 'junk' DNA that has no known purpose. Every time a cell divides, the entire DNA sequence must be accurately duplicated and copies given to each of the new 'daughter' cells. That makes good sense – the genes must go equally to both cells. If that doesn't happen, and there are occasional mistakes, one or both of the daughter cells will not have a full set of instructions at their disposal. With several vital paragraphs missing from the complete manual, and lacking the imagination to realize their loss or to improvise, the cells cannot function properly and they die. Even worse, if the missing pages contain genes which normally restrain cell multiplication, the cells begin to divide uncontrollably and may become malignant. Many cancers begin in this way.

But this is all recent knowledge, the fruits of modern-day technology applied by large and dedicated teams of research scientists. In the very earliest days of genetics, in the late nineteenth century, there were no clues at all about how genetic instructions might be passed on from one generation to the next. The pioneers of genetics realized that there must be some sort of message passing between parents and their children to account for the similarities between them, in appearance if nothing else. But what these instructions actually were and how they were transferred was a complete and utter mystery. At around the same time, biologists were starting to get tantalizing glimpses of the structures within individual cells. This was thanks to great improvements in the optical quality of microscope lenses and, in particular, to the use of new chemical dyes, developed for the textile industry, which could be used to stain different structures inside the cell with a palette of brilliant, strong colours. Without this treatment, the interior of cells was a colourless jumble of confusion; with them, structures inside the cell like the nucleus and the delicate tracery of the cytoplasm outside the nucleus could be seen for the very first time. When preparations containing dividing cells were dyed and placed under the microscope, strange threadlike structures could be seen within the cells. These were particularly strongly stained by the new dyes and were intensely coloured – which is how they got their name, 'chromosome' being derived from Greek for 'coloured bodies'.

Their function was unknown, but careful observations put together a common sequence of events. When cells were not dividing, the chromosomes were nowhere to be seen, becoming visible only when the cells were about to split in two. At first they looked like elongated threads but then, as the moment for division approached, they contracted and became much shorter: the stubby fingers I had seen under the microscope when I looked at my own chromosomes. Then an astonishing thing happened. The chromosomes lined up near the centre of the cell and each one was torn in two by muscular filaments anchored at each end of the cell. Half went one way and half went the other. (The cells that I had taken from my own blood never got this far because the colchicine in the growth medium had paralysed these filaments and so the chromosomes stayed put.) After the chromosomes had been pulled into opposite ends of the dividing cell, the cell itself split in two. Shortly after that, the chromosomes elongated once again and gradually faded from view.

But still nobody connected these strangely behaving objects with the transfer of genetic information. The early geneticists were more concerned with how genes were passed from one generation to the next than with the relatively mundane process of what happened when one cell divided to produce two identical 'daughter' cells. It was not until biologists saw chromosomes doing the same sort of thing in eggs and sperm that the penny finally dropped. The chromosomes and the genes were one and the same. From his breeding experiments with plants, the monk Gregor Mendel, working in what is now the Czech Republic, had deduced by the middle of the nineteenth century that pollen and eggs each had only one set of genes while adult plants had not one but *two* sets. He predicted that, when an egg was fertilized by a pollen grain, their two single gene sets must combine to reconstitute the double set of genes in the seed. When the seed grew into another adult plant, each cell retained the two complete sets of genes.

By the late nineteenth century, biologists studying the conveniently large eggs of sea urchins had actually seen chromosomes behaving in exactly that way – but, as they were unaware of Mendel's predictions, they did not make the connection between these strange threads and the secrets of inheritance. There was no doubt that the sea-urchin chromosomes were behaving exactly as Mendel had forecast. In the final cell division which produced the eggs themselves, the chromosomes were not ripped in two, as happens in normally dividing cells. Instead, each complete chromosome moved without fuss to one or the other end of the cell. Because the chromosomes had not split as the cell divided, each egg now contained only one set of genes. Though they could not see it, because the cells were so small, the same sort of division also preceded the production of sea-urchin sperm.

As the sun set on the 1800s and the twentieth century dawned, important pieces in the puzzle of genetics began to fall into place. Three independent scientists, each conducting his own plant-breeding experiments, were coming to very much the same conclusions as

Mendel had done forty years previously. Since then his published work had languished, virtually unread, gathering dust on library shelves. In the excitement surrounding the new plant experiments, Mendel's work was rescued from obscurity and he was immediately elevated to the eminence he now enjoys as the universally acknowledged father of genetics. Too bad for him that he never lived to enjoy his recognition. He had given up his experimental breeding to take on the burden of running the monastery, and died of kidney failure in 1884.

The confluence of Mendel's brilliant theoretical deductions and the clear views biologists now had of chromosomes and their peculiar behaviour under their microscopes very quickly crystallized into a new hypothesis – that chromosomes were the physical embodiment of Mendel's genes. The unexplained partition of chromosomes in the sea-urchin eggs became suddenly full of meaning. It was simply the process whereby the egg received its single set of instructions from the mother. Another single set of genetic instructions received from the similarly endowed sperm rebuilt a double set when the egg was fertilized. Thereafter, a long succession of straightforward cell divisions, starting with the fertilized egg, supplied two sets of chromosomes to all the cells of the body. Within a very short space of time during the 1900s, not only had geneticists confirmed Mendel's principles of inheritance in dozens of different species, both plant and animal, but chromosomes had also been found wherever dividing cells could be closely observed. At long last, it really looked as though a robust scientific basis for the mysteries of inheritance lay within reach.

Nevertheless, a lot of questions remained unanswered. Nobody knew how genes worked or exactly what the connection between genes and chromosomes really was. Chromosomes were being found everywhere and biologists feasted on the new treats available to anyone who could afford a decent microscope. Huge amounts of information flooded in from hundreds of different species and it became apparent at once that there were no fixed rules on how many chromosomes to expect. For sure, within a species all members had the same number of chromosomes; but there were big differences in the total number in different species, even between ones that were closely related. The numbers of chromosomes in a single set ranged from 4 in the tiny fruit fly to 7 in the peas Mendel used in his experiments to 15 in the lupin, 26 in the mouse and an astonishing 113 in some species of newt.

The first of these, the common fruit fly *Drosophila melanogaster*, soon emerged as a superb subject for genetic breeding experiments. These are the tiny insects that, in England anyway, you frequently disturb from your fruit bowl during the summer months and swat away without much concern. As soon as you look away, they're back on the fruit. Related species are found throughout the world doing just what the name suggests – eating ripe fruit. Fruit flies will live and breed almost anywhere. They are perfectly happy in an old milk bottle with some mashed-up banana in the bottom. They reproduce like wildfire with a breeding cycle of only ten days. Even nowadays, a hundred years after they were first bred experimentally, you can find a fly room in most university genetics departments, with the inevitable escapees flitting around unwashed mugs in the coffee room eager for a spilt drop of sweet liquid.

As well as being a low-maintenance, fast-breeding workhorse, the fruit fly had other advantages for the geneticist. Not all fruit flies look the same. There are scores of different features which vary among individual flies. There are flies with red eyes, flies with white eyes, flies with big wings, flies with small wings, flies with lots of bristles, flies with only a few bristles and so on. The list is endless. All of these features are ultimately controlled by genes which are passed on from one generation to the next in patterns of inheritance whose detail could be established by breeding experiments. The man who really capitalized on the potential of the fruit fly was the great geneticist Thomas Hunt Morgan. Active during the first three decades of the twentieth century,

Morgan was a strict disciplinarian, and supervised the world's first fly laboratory in Columbia University, New York, with an iron will. In the fly room, bench after bench was occupied by students each peering down a microscope at anaesthetized flies and methodically scoring them for a long list of characteristics. The doped flies were sorted into piles using tweezers and, if they were required for further breeding, released back into their milk bottles to recover and start life afresh with their assigned mating partners.

The amount of information on the fruit fly that came out of the Columbia fly room was immense. In the great majority of cases, the inheritance of the variable features – eye colour and so on – followed Mendel's rules precisely. However, occasionally it looked as though the rules were being bent. Only because these were such large-scale breeding experiments, with thousands of flies being studied, could the slight deviations from what was expected be picked up by the researchers. But it was the inferences drawn from these observed irregularities that were to clinch the precise relationship between genes and chromosomes that had thus far eluded everybody.

What the fly-room scientists first began to notice was that, from time to time, pairs of features were inherited together more often than they should be. This looked like a clear breach of Mendel's rule which set down that the inheritance of one feature was always completely independent of all others. But in the fruit fly this rule was being broken every now and again. Take as an example a milk-bottle mating between a fruit fly with the normal reddish eyes and short wings and a fly with a noticeably more brilliant eye colour (classified as vermilion) and long wings. Eye colour and wing length were being controlled by two separate genes. Slightly different versions of the eye-colour gene gave a fly red or vermilion eyes just as different versions of the wing-length gene produced long or short wings. If Mendel's law were being followed, you would expect equal numbers of red-eyed offspring with long wings and red-eyed offspring with short wings. But that isn't how it turned out. There were far more red-eyed, short-winged flies in the next generation than red-eyed offspring with long wings. What was happening was that the original combinations of features in the parent flies were being retained in the offspring more often than they should. Although the genetics is more complicated, it is comparable to the situation in humans where red hair and freckles go together.

The fly results made no sense until the chromosomes were brought into the picture. By degrees, the scientists in Morgan's team realized that, when they saw features being inherited together more often than expected, the genes which controlled them, the genetic instructions that gave the eyes their colour and the wings their shape, must both be contained *within the same chromosome*. This proved to be a brilliant deduction which was soon expanded to reveal the most astonishing behaviour of chromosomes and, most amazingly of all, the underlying reason for sex itself. Which is why I am telling you about it in such exhausting detail.

Once they realized what they had stumbled upon, the researchers at Columbia were alerted to the genes that bent the rules and were on the lookout for more. One of Morgan's most talented students, Arthur Sturtevant, soon discovered several pairs of features that followed the same disobedient inheritance pattern. He realized that the degree to which the features were retained in the offspring was different for different pairs of these mischievous genetic characters. For example, in the eye-colour/wing-shape case we took as an example, the combination in the parent fly stayed together in about 70 per cent of the offspring flies and fell apart in the other 30 per cent. Some of the new pairs of features which Sturtevant uncovered stayed together more often than this in the offspring, while other pairs remained together less often. But – and this was his absolutely crucial observation – however many times the experiment was repeated with a particular pair of characteristics, the percentages always remained the same. Eye colour and wing shape were always inherited together in 70 per cent of offspring, no matter how many times the experiment was repeated.

Bit by bit, the secrets of the chromosome were being teased out. Sturtevant couldn't explain the consistency of these numbers by concluding that the features stayed together simply because the two genes were on the same chromosome. There was more to it than that. He realized that the chromosomes themselves had to be far more fluid, far less permanent than they appeared to be under a microscope. If the chromosomes were fixed, then the red eye and short wing combination, and others like it, would be passed on intact to *all* of the offspring. But they were not. In 30 per cent of the offspring the combination was disrupted. Sturtevant gradually realized that the chromosomes, though they appeared as intact and continuous threads under the microscope, could be broken. If a chromosome broke between the two genes, then the combination of features they controlled would separate in the next generation. But if the chromosome remained intact between the two genes, then the combination of features would stay together.

Once he had made that intellectual leap, Sturtevant immediately realized why the percentages were different for different pairs of features. The rate at which the combinations were disrupted depended on *how far apart* their genes were on the chromosome. If the genes were a long way apart, the features would be separated in the offspring more often than if the genes were closer to one another. And since the percentages were the same for the same pairs of features no matter how many times he repeated the experiments, Sturtevant drew the far-reaching conclusion that the distance between genes on a chromosome was fixed. Not only that; on the basis of his breeding experiments he was able to put a figure on how far apart they were – not a strictly physical distance in fractions of a millimetre, but rather a genetic one, related to the chances of a chromosome breaking. In honour of the head of the fly lab, the unit of genetic distance was called the Morgan. The further apart two genes lay on a chromosome, the greater their genetic distance, measured in Morgans.

The realization that genes are arranged in a fixed linear order along chromosomes was a tremendous breakthrough. From the hundreds of thousands of genetic experiments in the fly room at Columbia a rational model for the relationship between genes and chromosomes finally emerged, a model which persists right through to the present day. It has led directly to the mapping of other genomes, including our own, and to the great triumphs of the last two decades in locating human genes at specific locations along our chromosomes. From that has followed the identification of the genes responsible for many of our most severe genetic diseases. Strange to think it all began in a crowded room in New York with old milk bottles, a few squashed bananas and a little fly that we hardly notice.

4

THE LAST EMBRACE

I have to admit at this point that I have been economical in my explanation of chromosome behaviour by not telling you the whole story behind what happens when chromosomes break. Though it is perfectly true that a chromosome break can occur between two genes, and that is what disrupts the combination going through to the next generation, what I did not say is that the break heals. But the truly astonishing part of it is that the healing process does not simply repair the original break; it joins together *two different* chromosomes.

As we saw earlier, animals – humans and fruit flies included – have two complete sets of chromosomes. For this reason, they, we are technically known as *diploid* (*two sets* in Greek). Some fish and amphibians and many plants have up to six sets of chromosomes, but we only have two. One set comes from our mothers via the egg; the other comes from our father's fertilizing sperm. Once the two sets of chromosomes find themselves in the same fertilized egg they divide when it divides and continue to do so throughout life, mechanically copying and splitting and generally minding their own business. In most of our body cells, our so-called *somatic* cells, the chromosomes that come from mother and the chromosomes that come from father, have very little to do with one another. Their genes carry on with their job of passing instructions to the cell, and the cell listens and obeys. It generally listens to the genes from both parents, because they are usually saying the same thing. Sometimes, in the case of genetically dominant features like brown eyes, one version of the instructions is preferred over the other. In all our cells, the conversations between our parents' genes continue through the chromosomes we have inherited from them, even when they themselves are long dead.

However, at a very early age in all of us, well before we are born, a few cells are set aside for a different purpose. These are called the *germline* cells, to distinguish them from the run-of-the-mill somatic cells which make up the rest of our bodies. These special cells are being groomed for the task of handing on the genes to the next generation. Once they have been selected they pursue a very different life from their somatic companions. While somatic cells will all eventually die, our germline cells can taste immortality. Though the details of development differ radically between men and women, the crucial genetic interplay of chromosomes remains the same for both sexes. After many rounds of cell division – far more in men than women, as we shall see later – germline cells reach the point when they must reduce their chromosomes from two sets down to one ready to package them into either eggs or sperm. This happens on the very last cell division. However, just before that last division something quite extraordinary occurs. The two sets of chromosomes, which have up to then led completely independent lives, come together for a final embrace.

Very gently the chromosomes find their opposite number and, starting at their extreme tips, they delicately lie alongside each other until they are entwined. Then the miracle occurs. Invisible breaks appear deep inside the touching arms. Very gently the cut end of one chromosome seeks out and joins with the break that has opened up in its partner. Healing enzymes close the wounds and the chromosomes begin to pull clear of their embrace. As they separate for the last time they linger for a final moment around the places where these intimate exchanges have taken place until, at last, they are pulled apart by the force of life and are parted for ever.

What can be the purpose of this strange liaison? Even though it lasts but a few moments, its impact on all of us is powerful beyond imagination. It is the very essence of sex itself. The chromosomes that emerge from the final embrace have changed their identity and their genes. These silent trysts have altered the chromosomes irrevocably. Before they touched and exchanged their gifts of DNA, they were all identical, exact copies of the chromosomes inherited from both parents. After the embrace is over, they are now mosaics of these chromosomes, part from one parent, part from the other. Because these exchanges occur randomly at more or less any point along the chromosomes, each mosaic is slightly different from all others. The new chromosomes all have a full set of genes, but the *versions* have been shuffled on each one, creating an almost limitless variety of combinations ready to pass on to the next generation. That is the reason we are all different. Identical twins apart, no two people have exactly the same genetic make-up. Thanks to our chromosomes' final embrace, brothers, sisters and non-identical twins never inherit the same combination of genes.

There was one last triumph to come from the Columbia fly lab. Having discovered that genes lie on chromosomes in a fixed order and that the chromosomes in germline cells break and rejoin at every generation to shuffle the genetic pack for their offspring, Morgan and a new arrival called Calvin Bridges made the connection between chromosomes and gender. The breakthrough followed earlier work in another laboratory, this time on a species of grasshopper with enormously long chromosomes that were very easy to see under the microscope. In the testes of male grasshoppers, one of these large chromosomes refused to join in the final dance that ended with the exchange of genes. It had no dancing partner. In the beautiful line drawings which record the strange antics of the chromosomes under the microscope lens in those early days – years before cameras could be effectively linked up – this mysterious chromosome was not given a number, like the others which come in pairs, but was labelled instead with the universal symbol of the unknown. It is marked on the drawings, in dark black ink, as chromosome X. That ambiguous accolade has survived from the drawings of the early microscopists of a century ago right down to the present day.

When the scientists from Columbia looked for the same chromosome in their fruit flies, they found it. They also found that, in females, it was not alone. It joined in the last dance with as much gusto as the others and swapped genes with its partner. Only in males was it alone. Or was it? In some of the best preparations of dividing cells from male flies there was a small chromosome, previously overlooked, which did behave as if it might just be the missing partner to chromosome X. For instance, it always ended up in a separate sperm from chromosome X, just as the individual members of the other pairs of chromosomes always went their separate ways after they left the dance. Then at last, in one cell, came the conclusive proof of this unlikely partnership. While the other chromosomes performed their intimate *pas de deux* and exchanged their genes, the end of chromosome X bent round to touch the tips of this tiny chromosome for the briefest moment. This was a kiss on the cheek compared to the long embrace of the other chromosomes – but it was proof of a relationship, however unlikely, however clandestine, between the two.

If the larger chromosome was X, what else could its unlikely partner be called but Y? At last, a genetic explanation for the fundamental difference between males and females was within reach. Females have two X-chromosomes; males have only one X-chromosome and one, much smaller, Y-chromosome. But that still left one question unanswered. Are the males male because they have a Y-chromosome or because they have only one X-chromosome? The answer came, as it so often does in science, through the discovery of cases that disobeyed the rules. Of all the qualities that distinguish a great researcher, the one I most admire is a talent for noticing when an observation or the result of an experiment just doesn't fit with the expectations. Fortunately, this talent was abundant in the Columbia fly lab, and the tremendous strides are attributable in large part to an almost intuitive feel for the rare exception. It was one of these exceptions that solved the puzzle. Just occasionally the breeding experiments produced females with the 'wrong' eye colour, according to the rules. Calvin Bridges was instructed by Morgan to investigate these apparent anomalies and, by examining the chromosomes of these exceptional flies, he worked out what was happening. He pinned it down to a mistake in the final cell division of the germline cells in their mothers. He found that *both* her X-chromosomes, instead of just one, had ended up in the same egg. This kind of mistake, which is given the name *nondysjunction*, can have serious consequences for humans; it is the cause, for example, of the extra chromosome in Down's syndrome. For Calvin Bridges' flies it led to a certain sexual confusion. He discovered that the unusual flies which had inherited two X-chromosomes from their mother had also received a Y-chromosome from their father. They were XXY flies. And they were female. Perfectly normal, fertile and fully functional females.

The same kind of chromosome nondysjunction in female flies also produced eggs with no X-chromosomes, a sort of mirror image of the double-X egg. When these were fertilized, by sperm containing an X-chromosome, they produced offspring with just one X-chromosome, but without a Y. And these flies were males. They looked perfectly normal but were in fact sterile. Bridges concluded correctly that the sex of fruit flies depends simply on the number of X-chromosomes. If you were a fly and you had two X-chromosomes you would be female. If you had only one X-chromosome then you would be a male. It didn't matter a great deal whether or not you had a Y-chromosome. Since humans had X and Y chromosomes too, everyone assumed that the same process decided sex for us as well. How wrong they were. It took decades for scientists to correct the mistake and to realize that for humans, the Y-chromosome was far from being the irrelevance it was in the fruit fly.

5

SEX AND THE SINGLE CHROMOSOME

It is very hard to believe nowadays, when we live in an age in which practically everything is given a genetic explanation, that unravelling the genetics of humans had an abysmally slow start and made only stuttering progress until comparatively recently. That is not to say there was never any interest in genetics among doctors. There certainly was, and as early as 1902 a few bright minds pointed out that some human diseases obeyed Mendel's inheritance rules and probably had a genetic origin. But without the facility for experimental cross-breeding, and having instead to rely on observations of naturally occurring 'experiments of nature', human and medical genetics had to wait until the DNA revolution of the early 1980s really to make its mark. This general lack of interest in genetics among most medical specialists, which I remember only too well myself as a young genetics lecturer, partly explains why it took so long to establish even the most basic of facts about ourselves – important things like how many chromosomes do we humans have?

Why this simple question should have taken so long to answer is something of a mystery. It is usually blamed on the considerable technical difficulty of counting chromosomes in thin slivers of tissue, which was all that was available at first. But I think it had more to do with the refusal of the few biologists who even bothered to look at human chromosomes at all to believe that there were diseases around that might have a chromosomal explanation. For example, it was left to an ophthalmologist, not a geneticist, to suggest that Down's syndrome might be caused by chromosome abnormalities. He had no way of proving this by himself and was reduced to imploring cell experts (cytologists) to investigate. That was in 1932. No-one took any notice, and it was another twenty-seven years before the extra chromosome that causes Down's syndrome was finally discovered.

Since chromosomes are really only visible through a microscope when they are condensed just before cells divide, it was hard to find enough cells anywhere in human tissues to do even the crudest of counts. The only tissue which had plenty of dividing cells was to be found in the testis, whose constantly busy cells, faced with the task of producing over 150 million sperm every day, are in a constant frenzy of division. But men do not gladly surrender their testicles, even in the cause of scientific research, so the early human cytologists were reduced to hanging around outside operating theatres or, even worse, waiting by the gallows for fresh specimens.

One of the most assiduous of the early cytologists, the Austrian Hans von Winiwater, pioneered the use of really fresh tissue and, in 1912, reported finding forty-seven chromosomes in males and forty-eight in females. Following the example of the fruit fly, he concluded from this that humans also had their sex decided by their number of X-chromosomes, with females having two X-chromosomes and men only one. What followed was one of those episodes that seem almost unbelievable when viewed with the great benefit of hindsight – an episode which compounded von Winiwater's error for almost forty years. In 1923, the American microscopist Theophilus S. Painter managed to get three testicles from the Texan State Lunatic Asylum which had been removed from inmates after 'excessive self-abuse coupled with certain phases of insanity'. He prepared thin sections of tissue and focused his microscope down onto the cells that were dividing. The chromosomes were there all right, but lying in a jumbled heap that made it very difficult to see where one ended and another began. This made counting them very hard but, after several months of indecision, Painter plumped for forty-eight as the correct number of human chromosomes. He did this despite the fact that, as we now know, in his clearest views under the microscope he counted not forty-eight but forty-six chromosomes. Exactly why he decided on forty-eight nobody knows, but the reason may have been no more rational than to keep von Winiwater company. In any event, the error completely blinded those who followed.

After Painter, no-one doubted that there were forty-eight human chromosomes until fully three decades later. Technical progress in making chromosome preparations was abysmally slow, and would have remained so had it not been for that most delicious of events – the accident observed. In the late 1940s a young Chinese graduate, T. C. Hsu, arrived at the University of Texas looking for a job. Hsu managed to find a position at the University of Texas Medical Branch at Galveston, Painter's old department, where his new boss wanted him to study human chromosomes working with the recently discovered science of cultured cells – cells grown in glass dishes from tiny pieces of human tissue. After a frustrating and fruitless six months during which he found that the chromosomes of cultured cells were just as jumbled and crowded as they were in tissue sections, the lab received a few samples of foetal tissue. These always grow well in culture, so Hsu set up as many different dishes of tissue cultures as he could.

Hsu decided to concentrate on culturing skin and spleen cells and, not expecting to see very much at all, given his previous six months' experience, almost casually stained a culture of spleen cells to see what the chromosomes looked like. He could scarcely believe his eyes. Instead of the usual jumble he was expecting, the chromosomes were scattered and beautifully separated from one another. He got up, walked round the building, had a cup of coffee and went back to his microscope. It had not been a dream. They were still there. He looked at more slides and all of them showed the same result. The chromosomes were cleanly dispersed over the slide, not lying in a tangled heap at the centre of the cell and impossible to count.

At once, he tried to repeat this with a fresh culture of spleen cells. When he examined these new preparations, to his horror he saw that the chromosomes had reverted to their original troublesome behaviour. There was no sign of the elegant spreads. He began to wonder if there had been something abnormal about the original spleen from which the cells had come, some peculiar pathology that had made them behave so miraculously. For the next three months Hsu repeated every step, trying desperately to remember whether he had done something different to the original culture. Then he began systematically to change the formulation of every one of the solutions that he had used on that wonderful day.

At last, he came to the salt solution which he had used to rinse the cells just before he put them onto the glass slides. When he diluted this solution with distilled water, the miracle reappeared. The chromosomes in these preparations were untangled and spread evenly across the glass, just as they had been that first day. He realized at once that there must have been an error in the formulation of the salt solution he had used that day. The technician whose job it was to make up these solutions must have made a mistake that had resulted in one bottle being more dilute than it should have been. No amount of enquiry could establish which of the young lady technicians was responsible. Whoever it was, she was naturally reluctant to admit to the error, even if she had been aware of it. So T. C. Hsu did not know whom to thank for helping him make the most important breakthrough in human chromosome research for over thirty years – and she remains an anonymous heroine to this day. No surprise, though, that Hsu's salary was raised and his papers for permanent residency filed at once by the university authorities. So proud were the university of their boy and his chromosomes that the dean of medicine, Chauncey D. Leake, was moved to verse:

> We think they may
> Fermenting quite unseen
> Direct and guide the symphony of life
> Which throbs eternally
> In every gene.

T. C. Hsu had capitalized on that rare occasion when an accident allowed a glimpse of the way ahead. But the most surprising aspect of his discovery was that, despite seeing slide after slide of beautifully separated chromosomes, he still added them up wrong. He believed so firmly that humans had forty-eight chromosomes, the number set in stone two decades earlier, that he never questioned it. So strongly was Hsu entranced by that particular spell that he refused to believe his own results when his counts disagreed with that magic number. It took a plant expert, who had not been raised among human geneticists, to break the spell.

Albert Levan, from the University of Lund in Sweden, was an experienced plant cytologist whose interests had turned to animal cells. He became fascinated by the similarity between the disorganized chromosomes seen in cancer cells and the changes deliberately induced, either by chemicals or by radiation, that he had seen during his research on plants. Levan got hold of some human embryonic lung tissues and, just as Hsu had done, set up a series of cultures which he then stained using the dilute salt pretreatment. But unlike Hsu, when he counted the chromosomes in his spreads he was not mesmerized by the 'spell of forty-eight' and consistently found only forty-six chromosomes in his preparations. Once he had published his findings, in 1956, the scales fell from the eyes of human cytologists everywhere and they soon confirmed Levan's result. At last, after more than thirty years, scientists now knew the correct number of human chromosomes. There were forty-six in the complete set, twenty-three from each parent.

The shockwaves that Levan's discovery set off at long last eclipsed the decades of complacency which had paralysed the study of human chromosomes. Suddenly doctors began seriously to consider chromosomes as a possible cause of the inherited diseases in patients under their care, and the comparative ease with which human chromosomes could now be studied using Hsu's treatment made investigations a practical possibility for the first time. One such pioneer was the French paediatrician Jérôme Lejeune, who had made a speciality of caring for children with Down's syndrome. These children are familiar to most of us, with their widely spaced eyes and often endearingly devoted and dependent behaviour. They are greatly loved by their parents, but the condition is serious, always associated with mental retardation and often with more sinister heart complications. Very few individuals with Down's syndrome live into their thirties.

Although an ophthalmologist had urged cytologists to look for chromosome abnormalities in Down's syndrome patients way back in 1932, none had taken the hint. But as soon as Lejeune heard one of Levan's colleagues talk about his forty-six chromosomes at a scientific meeting in Copenhagen, he decided to check on the chromosome numbers of his own Down's syndrome patients. The problem for him was that he had neither the training nor the facilities to do so; but, to his lasting credit, this did not put him off. He found someone in the hospital where he worked who did know how to culture cells and tissues and was willing to help. His own cramped laboratory had no running water for the staining steps of chromosome preparation, so he negotiated the use of a tiny adjoining kitchen. He didn't have a microscope, but he cajoled the bacteriology department into letting him have one of their cast-offs. It was so worn out that the cogs in the gears that adjusted the microscope plate on which the slides were placed had to be stuffed with silver foil from a bar of chocolate to prevent them from slipping. He didn't have a camera on his microscope, so he arranged with the pathology department to use their photographic equipment for two hours a week. Despite these hardships, Lejeune managed to grow cells from skin biopsies taken from his young patients.

The first of his patients to undergo the chromosome investigation was only two years old, and it is thanks to this little boy's bravery – skin biopsies are not painless – that Lejeune discovered the secret of Down's syndrome. When he treated the boy's cells with the dilute salt solution and then stained the preparation he counted not forty-six chromosomes but forty-seven. There was an extra chromosome. It was a very small one, certainly, but in size and shape, so far as could be told in those early days, it was not abnormal and could easily have matched up with either of a pair of normal chromosomes. We now know that the extra chromosome in Down's syndrome is number 21. Whereas normal children have only two copies of chromosome 21, Down's syndrome children have three copies of it. They have what is called a *trisomy* for chromosome 21. (All human chromosomes, except the X and the Y, are given numbers from 1, the largest, to 22, the smallest.)

When it was published in 1959 Lejeune's paper, though barely one page in length, had an immediate effect on the scientific community. Everyone at last woke up to the possibility that human diseases really could be caused by visible differences in the chromosomes, and in the few years that followed Lejeune's hard-won breakthrough many more inherited diseases were pinned down to chromosomal defects. Two other childhood diseases even more serious than Down's syndrome turned out to be caused by the presence of an extra chromosome, number 13 and number 18 respectively. Cytological examination of tissues from miscarried embryos showed that extra chromosomes were very often the cause of the premature termination of the pregnancy. Having an extra chromosome was clearly very dangerous. The impact of these and other discoveries was also rather disconcerting in a deeper sense.

Human chromosomes were not, after all, stable and reliable. They could fall apart or double up or do any of a multitude of other strange things and, what is more, do so at an alarming frequency. If chromosomes were the repository of our genetic blueprint, they were disturbingly fragile.

Despite the fact that the number of human chromosomes was now known for certain, no-one questioned that the way they determined sex was the same as in the fruit fly: two X-chromosomes in the female and one in the male with an irrelevant Y-chromosome tagging along but not doing anything. The next unlikely hero to clear the path towards the truth was a retired medical officer from the Royal Canadian Air Force. Murray Barr, originally a neurologist from the University of Western Ontario, had joined up as a medical officer and was based in England during the Second World War. Being interested in the structure of nerve cells, he had read about changes in their appearance under the microsope in, of all things, homing pigeons. During the dark foggy nights at the air base in East Anglia where he was stationed he wondered whether the same changes might be happening in the nerve cells of bomber pilots struggling to find their way home after the raids over Germany. Barr retained this particular curiosity and after the war, when he returned to his university life, he persuaded the Royal Canadian Air Force to give him a grant of $400 to follow it up – not in pilots but in cats, whose ability to find their way home had also impressed him.

As part of his research, Barr would routinely dissect out nerve cells and look at them under the microscope. As well as the changes to their overall shape, which was what he was most interested in, he also noticed that there was often a dark blob sitting within the nucleus. Whether the blob was there or not did not seem to depend on the experimental procedure to which the nerve cells had been exposed. There being no good explanation for this puzzling phenomenon, Barr dismissed it from his mind. Then, one night when he was working late in the lab, he had a good look through his records while he was waiting for an experiment to finish and noticed one very striking thing. The dark blob was found only in the nerve cells from female cats – never in the cells of males. Over the next few days, he looked at cells from other tissues and soon found out that the blob was not restricted to nerve cells alone but was there in any female cells he cared to examine. Before long Murray Barr had found the dark blob in many different mammals and, crucially for our story, in the cells of women too. But what was this dark blob that women had but men did not?

The final chapter in discovering how our sex is decided by the chromosomes came when Murray Barr examined the cells of a male patient with what is called Klinefelter syndrome. There is no doubt that these individuals are male, but they do have small testicles which are badly atrophied, with the result that Klinefelter patients are sterile. They also show body features that are more often associated with women and often develop small but definitely visible breasts. They grow little or no facial hair and are prone to osteoporosis in later life. And the cells of Dr Barr's Klinefelter patient contained the dark blob. Was it the presence of this blob – soon to be called the Barr body – that was responsible for the feminine features of this man?

The full explanation had to wait another ten years until two British cytologists, Pat Jacobs and John Strong, alerted by Lejeune's discovery of an extra chromosome in Down's syndrome, found that Klinefelter syndrome was also associated with an additional chromosome. In the cells of a 24-year-old patient they found not forty-six but forty-seven chromosomes. Since this discovery was made before the time when individual chromosomes could be positively identified, by the techniques I used on my own chromosomes as described in chapter 2, Jacobs and Strong could not be absolutely sure of the identity of the extra chromosome; but they correctly deduced that it was an extra copy of human chromosome X. This was the crucial discovery for the identification of the correct chromosomal mechanism for deciding sex in humans. The young man with Klinefelter syndrome had two X-chromosomes – but he also had a Y-chromosome. This meant that sex in humans could not depend, after all, on the number of X-chromosomes. If it did, the Klinefelter patient, with two X-chromosomes, would have been a woman. But he most definitely was not. For years, everyone had blindly assumed that humans followed the same rules as the fruit fly; but if this young man had been a fly he would, with two X-chromosomes, have been a female irrespective of whether or not he possessed a Y-chromosome. Suddenly the focus shifted onto the hitherto disregarded loner among our chromosomes, the Y-chromosome. Something on the Y-chromosome had prevented this individual from developing as a woman.

Confirmation of the central part played by the human Y-chromosome came soon afterwards from a different genetic disorder called Turner syndrome. Women with Turner syndrome are generally shorter than average with poorly developed breasts and only rudimentary ovaries. They also have just one X-chromosome instead of two – but they are indisputably female. Once again, this proved that humans were not following the fruit-fly rules, where having a single X-chromosome would make you male. It was obvious now that it was the presence or absence of a Y-chromosome, not the number of X-chromosomes, that determined whether a human embryo would grow into a boy or a girl.

Before long, perfectly normal women were being found with three or even four X-chromosomes, apparently without suffering any ill-effects. That was a surprise, considering how damaging it normally was to have even one extra chromosome, as in Down's syndrome. There must be something very special about the X-chromosome. The puzzle was partially solved when the cells of these women, and the patients with Turner syndrome, were stained for Barr bodies. They were nowhere to be seen in the women with Turner syndrome, but in the cells of women with extra X-chromosomes there were extra blobs. It didn't take researchers long to realize that the number of Barr bodies was always *one fewer* than the number of X-chromosomes. Earlier work on insects had occasionally described chromosomes that condensed down to dark-staining blobs and became inactive at the same time. Perhaps the Barr bodies were similarly shrunken X-chromosomes whose genes had been silenced. Since the number of Barr bodies was always one fewer than the number of X-chromosomes, it looked as if, in the normal human female, one X-chromosome was left active with its genes firing on all cylinders, while the others were closed down.

This made sense because it answered an anxiety that was growing among geneticists about the different number of X-chromosomes in men and women. The serious effects of having the wrong number of chromosomes, as in Down's syndrome and the other diseases where there were three copies of a chromosome rather than the usual two, made it very obvious that having the right number of copies of a chromosome was important for good health. It was as if the body needed just the right dosage of genes: neither too many nor too few. The problem was, if that were true, why did the fact that women had twice as many copies of X-chromosome genes as men not seem to matter? Surely this disparity between the sexes should inevitably lead to very serious abnormalities in one sex or the other?

The solution to this paradox is that in a woman one of the two X-chromosomes is permanently shut down, leaving only the genes on her remaining X-chromosome capable of doing anything. The inactive X-chromosome shrinks down to the dark blob first discovered by Murray Barr. In a man, on the other hand, the single X-chromosome is not shut down and is permanently on the go. The result is that the cells of both men and women are using the genes from only one X-chromosome. There is therefore no difference in the dosage of active X-chromosome genes between men and women; they are both making do with just one active X-chromosome. Men always get their single X-chromosome from their mother but women receive an X-chromosome from each of their parents. In men, no decision has to be made about which X-chromosome to shut down and which to keep active, but in the female embryo, at a very early stage of development, cells shut down either the X-chromosome from the mother or the X-chromosome from the father.

Once a cell in the embryo inactivates one or other of the X-chromosomes, it is permanent and irreversible. In every cell that is descended from it, the same X-chromosome is always shut down and, likewise, the other X-chromosome is always active. If you are a woman you are literally a mosaic made up of patches of cells in which one of your X-chromosomes is working while the other one is doing nothing, condensed down as a Barr body, and patches in which the other one remains active while the first one is closed down. This is impossible to see externally because, unlike some animals, we do not have genes on our X-chromosomes which affect features like skin or hair colour. But cats do, and the mottling on tortoiseshells, all of which are female, is produced by this mosaic effect. In most mammals, including humans, which X-chromosome is shut down in any particular embryonic cell seems to be completely random. It can be either the X they get from their mother or the one from their father. That's the case in most mammals; but in marsupials, like opossums and kangaroos, it is always the X-chromosome from the father that is silenced while the cells listen only to the chromosome inherited from the mother.

At long last, after forty years of confusion and misjudgement, the genetic essence of human sexuality – the essential genetic difference between men and women – was narrowed down to just one element: the Y-chromosome. If you've got one, you're going to be a man. If you haven't, then you're going to end up as a woman – it's as simple as that. But what exactly is it about this little chromosome that endows it with the power to decide between male and female, the deepest and most fundamental of all human differences and the cause of so much joy and pain, elation and suffering? Hitherto regarded as an irrelevance, this lonely outsider of the genome now found itself in the spotlight, and the search began to discover its intimate secrets and the source of its greatest strength. What had been seen as the least important chromosome of all was now revealed as holding the key to the gateway of human sexuality. But how did it do it? The first thing to do was find where on the Y-chromosome this power lay. The hunt was on for the sex gene.

6

HOW TO MAKE A MAN

In any hunt you need to know where to look for your quarry, and finding genes is no exception. The hunters knew the sex gene was lurking somewhere on the Y-chromosome – but where? Even though the Y-chromosome is unusual in many ways, drifting from one generation to the next with no partner, and is among the smallest of human chromosomes, it still retains the same overall structure as the others. All human chromosomes are divided into two parts, called its two *arms*, which are joined together at a structure called the *centromere*. The job of the centromere is to hold the chromosome arms together in dividing cells until the very last minute so that they do not fly everywhere. The centromere is also where the invisible filaments which pull chromosomes apart are attached. The all-important genes on a chromosome are strung out along its two arms on both sides of the centromere and, because the centromere is never exactly in the middle of a chromosome, one arm is always longer than the other. Sensibly, they are called the long and short arms.

There is a big difference in the length of the two arms of the human Y-chromosome and the long arm is usually about four times the length of the short arm (see figure 1). I say usually

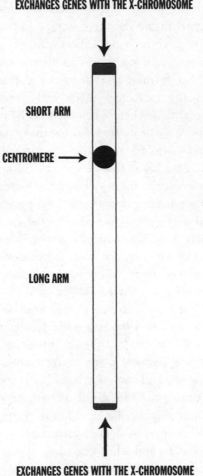

Figure 1: The Y-chromosome

because it turns out there are substantial differences in the overall length of the Y-chromosome long arm between individual men. Some men have much longer Y-chromosomes than others, and these differences are inherited. My Y-chromosome long arm, it turns out, is slightly bigger than the average and so, presumably, is Sir Richard's. The Y-chromosome short arm, on the other hand, is much less variable, and it also has another interesting property. Right at the far end, furthest away from the centromere, is a short piece which actually exchanges DNA with the X-chromosome during their brief embrace before cell division. This means the sex gene could not possibly be located right at the tip of the short arm because it could then be regularly transferred onto the X-chromosome. And we know that whatever causes maleness is not on the X-chromosome. So the very tip of the Y-chromosome short arm, and a tiny segment at the end of the long arm for the same reason, are ruled out as the hiding place of the sex gene. However, it could be anywhere on the remainder of the short arm or nearly the whole extent of the long arm. But is it necessary to have the whole Y-chromosome intact to produce a man, which would suggest that very many genes are involved, or is the crucial element concentrated in just a short stretch of the chromosome?

The first clues came in 1966 when Pat Jacobs, who with John Strong had been the first to find the additional X-chromosome in Klinefelter-syndrome patients a few years earlier, described two unusual cases that she had come across in her work as a cytologist in Edinburgh. Among the thousands of patients whose chromosomes she and her colleagues had examined, she encountered two unusual women. Neither had ever menstruated, and both had underdeveloped breasts and other secondary sexual characteristics, but otherwise they were unmistakably women, of normal height and intelligence. Examination of their cells showed that both women had only one X-chromosome instead of the normal two. However, their cells also contained another, very unusual chromosome. From its appearance – the pattern of dark and light bands revealed by the same Giemsa stain I used to identify my own chromosomes – it looked as though this was a Y-chromosome which, instead of a short arm, had a second long arm. This type of rearranged chromosome, called an *isochromosome*, is actually not all that uncommon among the other chromosomes, where they cause a range of symptoms almost always including mental retardation. Isochromosomes of the Y-chromosome had never been seen before and they obviously had not caused any mental impairment in these women. But the fact remained that they did have a large chunk of Y-chromosome – including a double helping of its long arm – and yet they were still women. That suggested that the sex gene could not be on the Y-chromosome's long arm. If it had been located there, these patients would not have been women, they would have been men instead. By a process of elimination, first the tip of the short arm, then the whole of the long arm of the Y-chromosome were ruled out. The field of search was narrowing – but getting any closer to the gene that creates men had to wait until the molecular revolution that swept through genetics in the 1970s and 1980s.

Genes are made of DNA and they sit on chromosomes. By a long succession of masterly technical breakthroughs, by the 1980s chromosomes could be dissected to their ultimate level – the sequence of their DNA. Tiny abnormalities in chromosomes, completely invisible even under the most powerful of microscopes, could now be picked up easily in a test tube. First segments of chromosome arms, then whole chromosomes were broken down into fragments small enough to have their DNA sequences completely read. Progress on a scale unimaginable even ten years before opened up the entire human genome to the most detailed analysis, and within a very short space of time the chromosomes began to reveal their secrets. They were no longer the enigmatic phantoms that appeared and disappeared during the life cycle of a cell. Like an unexplored inner continent they were mapped, first by finding fixed points of reference and then by using these to triangulate the rest of the landscape. The genes for major inherited diseases, cystic fibrosis, the muscular dystrophies and several forms of inherited cancer among them, were first located to specific chromosomes and then tracked down and their DNA sequence read through. The mutations in DNA that cause these dreadful diseases were found and rapid tests soon developed for their diagnosis.

It was a wonderfully exciting time which I remember very well. Almost every week the scientific journals, and often the popular press, announced the discovery of new disease genes. Competition to find them was intense and well publicized. The races between leading research groups drew all of us into this spectator sport where the prize for winning was the glory of being first, and where the losers got nothing.

Though the sex gene was not generally thought of as causing a disease, the race to find it was similarly frantic. The earlier work by Pat Jacobs with her two women patients had narrowed its location down to the short arm of the Y-chromosome. Though only a tiny island under the microscope, this was still a vast continent on the molecular scale, at least twelve million DNA bases long. The haystack might have been identified; the needle remained hidden. How could scientists get the gene within range of their molecular artillery, the array of new techniques which could finally nail the gene and discover its DNA sequence? At the time, in the late 1980s, the location of the target had to be accurate to within a few hundred thousand DNA bases. In a chromosome arm of twelve million bases, the geneticists had to know where to direct their fire.

Not for the first time, it was the unusual patient that proved to be the decisive factor. The first was a woman with only one X-chromosome, but also a Y-chromosome short arm transplanted onto one of her other chromosomes – number 22. These kinds of changes, called *translocations*, where parts of one chromosome break away and attach themselves to another are, perhaps surprisingly, relatively common. As long as the transfer is complete and no genes are lost or broken in two, translocations can be completely harmless to the individuals who carry them. By and large most genes don't mind which chromosome they are on. Problems arise only if, because of the translocation, their children receive the wrong set of genes, either too many or too few.

In 1986, the patient with the Y:22 translocation came to the attention of David Page, a scientist working at the Whitehead Institute in Cambridge, Massachusetts, one of the foremost biological research institutes in the world. Page had been interested in the Y-chromosome for a long time and had already built up a complete collection of DNA fragments, each cloned into bacteria, that covered the entire short arm of the chromosome. Without wanting to stretch the metaphor too far, you could say he had split up the haystack into several hundred bales. Using these fragments, Page was able to check whether a Y-chromosome was complete and intact, or whether there were small sections missing, just by testing its DNA.

When he checked through the DNA of this female patient, Page discovered that the Y-chromosome which had attached itself to her chromosome number 22 was not complete. This was as he expected: for if the entire short arm had been transferred, Page argued, it would have carried the sex gene with it and the patient would have been a man. The translocation was missing a small segment of Y-chromosome which was 160,000 DNA bases long. This was certainly a very large chunk of DNA, but still only about 1 per cent of the DNA in the short arm. Since Page's DNA checks established that the rest of her Y-chromosome short arm had been translocated intact, this one patient narrowed the search for the sex gene down to this comparatively small segment of DNA, within the range of the DNA sequencing skills of the time.

Page's confidence that he was closing in on the sex gene was boosted when he found a second anomalous patient – a man who, rather than having the normal XY male package, had two X-chromosomes instead and so should, by rights, have been a woman. Under the microscope his chromosomes looked perfectly normal, with no visible sign of a Y-chromosome or any parts of it. But when David Page checked the man's DNA against his test kit of Y-chromosome fragments he found that he had the same segment of the chromosome, the same bale from the haystack, that had been missing in the woman with the Y-chromosome translocation. In this man it was lodged somewhere on one of the other chromosomes and was just too small to see by eye.

This was a wonderful combination of two completely independent sets of circumstantial evidence. First, a female patient with a Y-chromosome short arm lacking just a comparatively small segment of DNA; then another patient, this time a man, with barely any of the Y-chromosome – except for the segment that was missing in the woman patient. Page had shown beyond doubt that it was unnecessary to have a complete Y-chromosome to become a man. And, most important of all, he had narrowed down the search for the sex gene to a tiny piece of the Y-chromosome, well within range of his molecular arsenal. He was getting very close.

After a few months of sustained bombardment of this crucial stretch of the Y-chromosome with the latest tools of the genetic engineering trade, David Page found a gene and gave it the codename DP1007. I am sure I am not alone in noticing the initials and a certain masculine resonance in the last three digits. Very quickly, Page and his team read through the DNA sequence of DP1007 and used that information to find out what sort of gene they were dealing with. As we saw in chapter 3, DNA is a long linear code which instructs cells how to make proteins, and cells read these instructions to decide the order of amino-acids in the proteins they build. By reading a gene the cell learns which amino-acids to use, and in which order, when fashioning the corresponding protein. Scientists can do the same. It is very straightforward for a scientist to work out the order of amino-acids in the protein that any gene specifies by reading through a DNA sequence. It is the amino-acid sequence of a protein that decides its function and, because of this, proteins which do similar things often have similar amino-acid sequences. So, if you have found a new gene, as David Page hoped he had with DP1007, comparing the amino-acid sequence of the protein it encodes to the sequences of known proteins can give a big clue as to what it might be doing in the cell. When David Page did the comparisons with DP1007 the result could not have been more compelling.

The gene that Page had discovered in just the right place on the Y-chromosome contained the DNA instructions to build a protein which bore a remarkable similarity to a family of proteins that were already well known to scientists. They were called *transcription factors*, and their job was to act as molecular switching devices – to switch other genes on and off. This was almost too good to be true. Nobody had ever seriously imagined that all the things that made a man could be contained within a single gene. If sex was decided by just a single gene, as looked increasingly likely, then it had to be some sort of master switch; a switch that, once flicked on, would activate all the downstream processes required to build a man.

The amino-acid comparisons that Page and his colleagues had run against DP1007 identified it as a molecular switch by virtue of a molecular structure within it with the medieval-sounding name of a 'zinc finger', a structure shared by other transcription factor molecular switches. How appropriate that the male master switch should take the form of a knight's gauntlet, the galvanized digit pointing the way to a life of chivalry and adventure. In fact, the zinc finger has a much more prosaic etymology. It is so called because of its molecular shape: a bit of it sticks out and binds an atom of zinc. Nonetheless the new gene, first codenamed DP1007, was now christened ZFY – short for 'zinc finger on the Y-chromosome'. Combining all this evidence with his discovery of a similar gene in the right place on the Y-chromosomes of mice, David Page felt confident enough to announce his discovery to the world and it appeared, in record time, in the 1987 Christmas Eve edition of the well-known US science journal *Cell*. You can just imagine how the other research teams throughout the world, racing for the same prize, choked on their roast potatoes the following day. To lose the race for any gene is bad enough, but to be pushed into second place in the search for the essence of maleness is particularly galling – especially if you are a man. It no longer looked like a coincidence that David Page had christened the crucial fragment DP1007, combining his own initials with a licence to kill. But the triumph, like so many of James Bond's conquests, was short-lived.

The initial reaction among scientists to the publication in *Cell* was enthusiastic. It really did look as though the hunt for the vital element that fundamentally distinguishes men from women had at last reached a successful conclusion. The answer appeared simple and very elegant. A single gene on the Y-chromosome, when activated, switched on a cascade of other genes, as yet unknown, that diverted the embryo from its natural course of development into a female and instead channelled it along a different path, the path that leads to maleness.

But even as the applause was dying down, slight cracks began to appear in the apparently watertight case for ZFY. One was that a matching gene was found on the X-chromosome. This observation, contained in Page's paper, was not at first seen as an insurmountable obstacle, but it did suggest that some adjustments might be necessary. After all, the careful work on chromosomes had firmly concluded that a single gene on the Y-chromosome was both necessary and sufficient for the development of maleness. The DNA-matching test could not tell whether the corresponding gene on the X-chromosome, called ZFX, was active or not. One possible explanation was that it had experienced a mutation which put it out of business at some time in the past and it had lingered on as what is called a *pseudogene*, a ghost gene that is still there but is fatally wounded by mutation and no longer able to function. The human genome contains a great number of pseudogenes, drifting through time without a purpose. To suggest that the copy of the

X-chromosome was similarly disabled was not at all far-fetched.

ZFY's short but dazzling career as the sex gene was fatally wounded by the discovery that in wallabies and other marsupials it was not located on the Y chromosome at all, but elsewhere among the other chromosomes not involved in deciding sex. The conclusion was a clear choice. Either marsupials used a different system altogether to determine sex, which was unlikely, or ZFY was not the sex gene after all. Maybe the race was not yet over and the gold medal might have to be returned.

Meanwhile, a few XX men who were missing the ZFY gene were being discovered in other research centres. These patients did, however, each possess a very short segment of the Y-chromosome from even closer to the tip of the short arm than ZFY, very close to the boundary with the part of the Y-chromosome which exchanged DNA with the X-chromosome. Could it be that the sex gene was squashed right up against this boundary, so close that no-one had seriously thought of looking there? When David Page announced that the XY female so crucial to his own discovery of ZFY was also missing a short segment of DNA very close to this boundary, the pace of research quickened even more.

The rival research teams, their hopes revived by the demise of ZFY, threw all they had at this segment of DNA, and it was not very long before another gene was found. When decoded, the gene made a protein that was strikingly similar to a protein known to switch genes on and off in yeast. Just like ZFY, the new gene had the potential to be the sex master switch and, as a statement of confidence, it was given the name SRY – short for Sex-determining Region on the Y-chromosome. Even more encouraging, unlike ZFY, there was absolutely no sign of a matching gene on the X-chromosome. It all looked very convincing when the discovery was announced in the journal *Nature* in July 1990.

SRY might show all the qualities expected of the sex gene, but was it alone sufficient to alter the course of embryonic development from female to male? The answer came the following year in

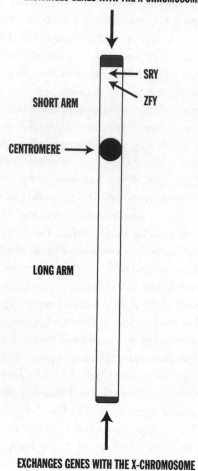

Figure 2: The Y-chromosome and the search for the sex gene

a conclusive experiment which silenced any doubters once and for all. The team led by the British geneticists Peter Goodfellow and Robin Lovell-Badge, who had been the first to find SRY, injected fertilized mouse eggs with a small fragment of DNA which contained the SRY gene and nothing else. No junk, no other genes, just SRY. They re-implanted the eggs back into female mice, which acted as surrogate mothers, and waited for the babies to be born.

These gene-transfer experiments are notoriously inefficient. The injected gene has to find a home on one of the mouse chromosomes to survive, and there was no guarantee that the SRY gene had successfully done this. Peter and Robin were looking for mice which looked like males but had two X-chromosomes *and* the injected SRY gene. Out of ninety-three mice that were born, only one had this combination – but it looked and behaved like a perfectly normal male. The sex in that one mouse had been reversed from female to male by the SRY gene alone. And one was enough to prove the point. There was nothing else from the Y-chromosome in this mouse. It was indeed a triumph and their star mouse, swinging on a stick and sporting enormous testicles to prove the point, made the cover of the edition of *Nature* that carried their article. Sadly, male mice with two X-chromosomes are always sterile so he could not have any offspring. That didn't stop him trying, though. When caged with female mice for company he mated four times in six days – a good average for a mouse, apparently. This dramatic demonstration of sex reversal, where a female embryo had been turned into a male using nothing more than SRY, finally closed the last chapter in the long search for the master switch that created men. From the moment when scientists realized that it was the Y-chromosome that held the secret to the final unmasking of the gene itself, the quest had taken thirty long years.

Though this book is about the genetics rather than the anatomy of sex, it would be churlish to abandon the explanation just at the point when the master gene is discovered without saying anything at all about how it works. Sadly – or perhaps fortunately, if you are already feeling overwhelmed by detail – not a great deal has been found out about the precise way in which the master SRY gene does its job. Like so much in life, it has proved easier to find than to understand. SRY clearly has the ability to switch on other genes on distant chromosomes, and no-one pretends that it acts alone. Exactly how these other genes work and in what order they become activated is still shrouded in uncertainty; still, the anatomical results are plain to see.

For the first six weeks of development, human embryos destined to become male and female are indistinguishable from one another. We know, of course, that one has two X-chromosomes and the other an X- and a Y-chromosome, but up to this stage of development there is no way, short of a genetic test, of telling them apart. They both have a pair of unisex gonads and two sets of primitive tubing called the Wolffian and Mullerian duct systems, named after their eponymous discoverers. During the seventh week of gestation the master gene, embedded in the Y-chromosome, is switched on in the male – but only for a few hours. The SRY protein, built to the precise orders of the sex gene, peels off the production line and heads off to activate other genes on several different chromosomes. From there, these genes trip a succession of genetic relays and under the influence of these secondarily activated genes, his unisex gonads begin to develop into testes which, before long, start to produce two different hormones. One is the descriptively named anti-Mullerian hormone or AMH, which effectively destroys the Mullerian duct system.

The other hormone produced by the embryonic testis is much better known. It is testosterone. At this early stage in the growing male embryo, testosterone prevents the other system of primitive tubes, the Wolffian ducts, from being destroyed as they are in women. As time passes the Mullerian ducts disappear and the Wolffian ducts begin to expand to form the components of the internal male sexual organs – the prostate gland and seminal vesicles, and the *vas deferens* which connects them. Finally, some of the testosterone is converted into a high-octane form of the hormone – called dihydrotestosterone – and this organizes the growth of the external genitalia. Folds of tissue surround the urethra and form the penis, while nearby other tissues swell and fuse together to become the scrotum into which the testes eventually descend.

Female embryos, oblivious to the genetic stirrings on the Y-chromosome because they don't have one, proceed along their developmental pathway undisturbed by the irresistible hormonal signals coursing through their male counterparts. At about the twelfth week of gestation the unisex gonads begin to transform into ovaries. The Wolffian ducts, unsustained by testosterone, fade away and the Mullerian ducts, unsuppressed by the combative AMH hormone and encouraged by oestrogen, begin to form the female ducts. The forward parts form the Fallopian tubes while the remainder develop into the uterus and the vagina. On the outside the same tissues which in the male develop into the penis form the clitoris, while the surrounding tissues swell and go on to become the major and minor labia rather than the scrotum. All of these anatomical changes are in place by the twentieth week of pregnancy, when the sex of the unborn child is visible on an ultrasound scan. Twenty weeks later, they are out, their anatomical labels read and announced to the world.

Now we know what it takes to build a man. The Y-chromosome works hard to prevent men from turning into women and we now have a pretty good idea of how it does it. But as one question is solved, so others beg to be answered. Why go to all that trouble to create two sexes in the first place? Indeed, why bother with sex at all?

7

SEX TIPS FROM FISH

You could be forgiven for thinking that most other species that indulge in sex would create their males and females along similar lines to ourselves. But nothing could be further from the truth. While sex is almost universal, the ways of deciding how to organize it are anything but, and the methods adopted by different species come in a bewildering variety of disguises.

Beneath the white surf and the crashing waves of any Pacific coral reef, the seawater surges to and fro with rhythmic ease. Close to the coral walls, companies of fish sit swaying in the current, darting out to capture scraps of food or suck closely at the algae on the rocks. Bright yellow tangs swim in close circles round and round the antlers of a stag coral. Turquoise parrotfish, their mouths hardened into beaks to tear off pieces of coral, move in fast straight lines like commuters late for work. A dozen different butterfly fish, each embossed with an impossibly intricate design, move sedately in the swell, picking delicately at the rocks like models at a cocktail party. Among these ravishing beauties of the reef you could easily overlook the small blue-headed wrasse as it swims among the coral. But it is a very handsome fish in its own right, its violet head brushed with broad yellow stripes that taper down its streamlined body. Observing this little fish darting to and fro between the coral heads, it is hard to realize what a bizarre sex life it leads. Blue-headed wrasse live in harems with a dozen or so females being jealously guarded by a single male. That in itself is not unusual – it's a popular arrangement even among humans – but what is so very strange about the blue-headed wrasse, and very clever too, is that it can change sex virtually at will. When the colourful male dies, or is removed by a spoilsport researcher, the largest female in the harem, and she alone, begins to change colour and to adopt the flashy outfit of her now departed lover. She is literally changing into a male. This transformation takes about a week, after which she actually becomes a male in both appearance and in behaviour. From then on he/she runs the harem, in time fertilizing, with the sperm he/she is now able to manufacture, the eggs of her one-time female companions.

The blue-headed wrasse has abandoned chromosomes as a way of deciding sex and instead relies on a purely social signal, the disappearance of the male from the group, to govern the process. Other animals have even weirder methods. Take one of my own personal favourites – the marine worm *Bonellia viridis*. This visually unappealing creature spends all its time in a burrow under the mud of mangrove swamps in the warm waters around Malaysia and Indonesia. To feed, the worm pushes out an enormously long proboscis, almost a metre in length, which sweeps to and fro in a broad swathe around the mouth of its burrow, picking up food. All these worms are females; the males are nowhere to be seen. That is because they actually live inside the females. Despite the metre-long 'tongue', the body of the female is only 8–10 centimetres long. But that is enormous compared to the tiny male, who is only 5 millimetres long. This much-reduced creature lives inside the female's womb, where he feeds on her nutrients and needs only to produce sperm when she is ready to lay eggs. Imagine having a husband like that, the ultimate sperm-delivery system, tucked away out of sight and reduced to a single function – producing sperm to fertilize your eggs.

That is fantastic enough, a masterpiece of efficiency and perhaps the ultimate in male subjugation – or laziness if you prefer. But it is matched by the devastating ingenuity of the way in which the sex of the worm is decided in the first place. Young *Bonellia* pass through a larval stage where they wander about in the mud. At this point they are neither male nor female and have the potential to develop into either sex. When the time arrives for them to change into adults they settle down on the surface of the mud. If a young worm's chosen spot is within the arc of a female's sweeping tongue, a hormone secreted by the ranging proboscis decides its sexual destiny. Once touched by this wandering tongue and intoxicated by the hormone, the larva inches helplessly towards the female, enters her womb and takes up residence. Within a few short weeks, all the necessary organs develop and the enslaved male starts pumping out sperm. Larvae that settle beyond the reach of a female's tongue stay where they are and themselves grow into females who, once mature, begin casting about, quite literally, for a mate.

There is nothing chromosomal about sex in the marine worm or the blue-headed wrasse, and there are plenty of other examples of sex being decided not by an intrinsically genetic mechanism but by an external stimulus. Take turtles, for example. When the time comes to lay their eggs, the great marine turtles haul themselves out onto their breeding beaches under cover of darkness, just as their ancestors have done for millions of years before them. The female strains to heave her body, heavy with eggs, up the sloping sand until she reaches a point above the high-tide mark where it is safe to dig a nest. This done, and the eggs having been laid and covered with sand, she turns once more to the sea and, going more easily downhill, slips into the waves for another year.

In the nest, the eggs as laid are neither male nor female. The sex of the hatchlings is not a matter for their chromosomes but is left to the temperature of the sand. If the sand is too hot or too cold the eggs will not hatch at all. Only if the eggs are incubated somewhere between 26°C and 34°C do they develop into young turtles. But within that eight-degree range there is scope for deciding the sex of the babies. If the temperature stays at the hot end then they will all be females, while if it stays at the cool end, the brood will be entirely male. Only in the middle of the range, around 30°C, will there be roughly equal numbers of both sexes. Whether the vulnerable hatchlings that run the gauntlet of gulls, skuas and other predators as they head valiantly for the safety of the surf are male or female depends entirely on the temperature of the sand in which their nest lay hidden.

It sounds almost careless to leave such an important decision as the sex of your offspring to the vagaries of the weather, but turtles are not the only animals to let the heat decide. Mississippi alligators do it as well. However, in alligators, the turtle rules are reversed. Cool eggs turn into female alligators, warm eggs into males. Unlike turtles, which all make nests in much the same place above high-water mark, alligators lay their eggs in different locations which experience different temperatures. In the Louisiana swamps, high up on the riverbanks, the nests are dry and warm, and the baby alligators hatching from eggs laid in these warm places are all males. By contrast, eggs laid lower down in the wet marshes experience a cooler incubation and they all hatch as females. Only in the intermediate locations, above the waterline but not too high up, do both sexes hatch from the same brood. Compared to the turtles, the Mississippi alligators are able to exercise a degree of control over the sex of their offspring by choosing where they lay their eggs. Also, unlike the turtles, they do not abandon their eggs after laying them to head out to sea for another year. Nevertheless, both alligator and turtle are very vulnerable to sudden and sustained changes in temperature.

Could it be, as a few scientists have suggested, that the dinosaurs, close reptilian relatives of both alligators and turtles, also used this same method of deciding the sex of their offspring? If that were the case, might this have been the immediate cause of their extinction? If the giant meteorite that crashed into the earth 65 million years ago reduced the temperature by those crucial few degrees, it may have been its specific effect on sex determination rather than a more general influence on the food supply that exterminated the dinosaurs with such rapidity. If dinosaurs also used the cool = male system adopted by the turtles, a sustained drop in temperature over several seasons would have meant a huge drop in the number of females being born – and no species can survive without females. Early mammals and birds, relying instead on a less vulnerable chromosomal mechanism for sex choice, would not have been affected in the same way, and could have lived through the cold years.

Before returning to the familiar territory of our own species, let us pause to look at one of the most versatile ways of choosing sex, the one developed by many different insects including the familiar ants, bees and wasps which live in colonies. Whereas in most insects, like the fruit fly we have encountered already, sex is decided by how many X-chromosomes you have, bees have taken this one stage further. Instead of being decided by the possession of either one or two copies of a single chromosome, in bees, wasps and ants sex depends on having either one or two *complete sets* of chromosomes.

Take the familiar honey-bee. The only bees you are ever likely to see in the open, the ones which buzz from flower to flower in your garden, and will sting you if you disturb them, are females. They are the workers and they have two sets of chromosomes, just as we do. They, like us, are *diploid*. Just like us, they inherit one set of chromosomes from their mother, the queen, and one from their father, a drone. The queen is obviously also a female and, like the workers, has two sets of chromosomes. She is the only female in the hive to be fertile, the only one that can lay eggs. However, the twenty or so drones in a hive, the males, do not have two sets of chromosomes; they have only one. This situation is known as *haploid* – 'one set' in Greek.

The drones inherit their single set of chromosomes from the queen. But, unlike the workers, the drones get no chromosomes at all from a father. They, quite literally, don't have a father. The hive is full of workers and a few drones. The queen is the mother of them all; she lays all the eggs. Each of the workers has a drone for a father but the drones themselves do not have a father. How does the queen manage to lay two sorts of egg – those containing two chromosome sets, which hatch as females, and those with only one set, which hatch as males? The answer is absolutely ingenious. When the drones mate with the queen, she stores their sperm in a special organ, a small cavity, called the *spermatheca*. This sits conveniently alongside the tube through which her eggs pass out of her body as she lays them one at a time. In a way, she is able to decide the sex of her offspring. When she chooses to lay in a standard cell her body is slightly squeezed, and sperm are released to fertilize the passing eggs which then hatch as diploid females. However, if she lays an egg in a slightly larger 'drone' cell, the sperm remain in storage, the egg is not fertilized and is laid with only her set of chromosomes and so hatches as a haploid male. It is a magnificently versatile system and one which allows mothers to choose the sex of their offspring to suit the conditions.

These are just a few of the strange ways sex can be decided – the chromosome system widely adopted, though with modification, by species as different from each other as fruit flies and humans; the risky dependence on egg incubation temperature favoured by turtles and alligators; the elegant simplicity and great flexibility of the haploid/diploid mechanism in bees and other social insects; the extremely practical sex changes in the blue-headed wrasse and the sinister entrapment tactics of the marine worm. They illustrate the universality of sex and the multiplicity of ways to achieve it. These examples all show that sex can be decided in many different ways, but they do not answer an even more funda-mental question. Why have sex in the first place?

8

WHY BOTHER WITH SEX?

At first glance, the answer to the question why we need sex is completely obvious. We imagine that without sex there would be no breeding, no offspring and no next generation. Surely, without sex, all animals would become extinct. In one very obvious sense, that is certainly true. If we never had sex, we wouldn't have any children. The answer sounds so obvious that the question itself seems absurd. But is it? Certainly, as things stand now, we humans cannot reproduce without some sort of sex, though actual intercourse is no longer compulsory. For the last twenty-five years, since the introduction of in-vitro fertilization, eggs can be mixed with sperm in a test tube without the donors ever having to meet in person. But that still counts as sex: the deliberate bringing together of DNA from two individuals, one female and one male.

Let us first challenge the natural assumption that sex is essential for reproduction. In fact, sex and reproduction have quite opposite purposes. One cell divides to become two – that's reproduction. Two cells fuse to become one – that's sex. Although we naturally combine the two processes into one and call it sexual reproduction, they are not, at the very basic level, quite so dependent on each other as we might at first imagine. In some species, it turns out, reproduction is perfectly feasible without sex. Since for survival, reproduction is biologically more important than sex, asking why we and almost all other species bother to involve sex in the process is a perfectly fair question. And it's one that scientists have struggled for a long time to answer.

Consider the greenfly that cover the stems of garden plants in the summer and suck out their juices. They manage to reproduce very well indeed without much sex. For most of the summer, repro-duction for an aphid is just a matter of giving birth to exact replicas of itself. They do this at an alarming rate, as you will know only too well if you are a gardener. With a magnifying glass you can see the small green clones being extruded from the mother's back end as she carries on feeding with the front. Even as these young aphids emerge they already contain the embryos of their own children. This is cloning on a grand scale – the creation of exact genetic copies one after another. However, towards the end of the summer, aphids do indulge in a little bit of sex. Instead of endless generations of cloned females, a few males are born. They mate with the abundant females, who then produce their own clones – new offspring with new genetic combinations. Even though the greenflies do have sex at the end of the season, it is obviously not essential for reproduction. For most of the year they manage perfectly well without it. But there are also some creatures that have abandoned sex altogether.

One celebrated example of sex-free reproduction is the whiptail lizard (*Cnemidophorus uniparens*) found in the south-western United States. These successful reptiles seem to have worked out a way of completely dispensing with sex. No male whiptails have ever been found and the single sex, which we could call female for convenience, simply lays eggs which hatch as daughters which are genetically identical to their parent. Just like the summer greenfly, they have evolved a very simple and very straightforward way of reproducing which, on the face of it, looks extremely appealing from the point of view of efficiency. The species is doing well; the lizards are perfectly healthy, all of them descended from a single female that discovered how to reproduce without a male. The whiptail lizard is probably the most complex animal to reproduce without sex in the wild but, among animals, there are plenty of insects, and many species of fish, which have also adopted cloning as a way of life.

Sex-free reproduction is even commoner among plants – many species of strawberry, blackberry and dandelion reproduce this way – and among much simpler organisms reproduction without sex is even more frequent. There is even an entire family of tiny microscopic animals, with at least five hundred different species, that have abandoned sex altogether – that is, if they ever knew what it was in the first place. These are an unassuming form of pond life called *bdelloid rotifers*. Different species thrive in any sort of fresh water, from geysers and hot springs to the pools of water that form on the surface of Antarctic ice in the summer. In a glass of pond water, they are just visible to the naked eye as pale moving specks; under a microscope you can see these tiny creatures zipping through the water and sweeping up anything in their path into their mouths with rapidly beating hairs. When the rotifers sense that their puddle is about to dry up or freeze over, they transform themselves into armour-plated spores, called *tuns*, that are virtually indestructible. These tuns are blown into the atmosphere and travel as dust, sometimes for thousands of miles, before coming back down to earth, where they lie dormant in the soil. When it rains, the creatures burst free from their drought-resistant capsules and begin to feed and reproduce all over again. In no time the puddle is swarming with rotifers, all identical cloned descendants of the incredible travelling tun. And all without a whiff of sex. Sex just cannot be essential for reproduction. Sex-free reproduction is going on all around us, all the time, and, on the face of it, does seem to have a lot going for it. Less fun perhaps, but with plenty of compensations. And one of them is efficiency.

Imagine for a moment that we are designing an entirely new

species of animal, and among our tasks is to decide on its method of reproduction. Should we choose the familiar arrangement of males and females reproducing with sex, or a sex-free scheme with just one gender which reproduces by cloning? To simplify matters, we shall abandon any notion that sex is fun, which it certainly isn't for most animals; and, so that we can focus on something, let us suppose we are creating a new species of rabbit. The objective is to make our new rabbit as successful a species as possible, and we define success simply in terms of surviving numbers. To make it a fair comparison we will design not one but two new species of rabbit that are identical in every respect apart from their method of reproduction. We will give them identical appetites, identical life-spans and identical periods of gestation.

How shall we compare their success? Let us imagine that we have enclosed a meadow with enough grass to feed a thousand rabbits and plenty of space for burrows. Now let us introduce two rabbits from each of our new species and measure their comparative success by seeing how many of each there are by the time the field is full. Let's call the two species Virgin and Regular. Virgin, you've guessed it, reproduces without sex, while Regular rabbits are restricted to the conventional method. For simplicity, let's say the rabbits produce one litter of four young after three months and then die. So there are four generations every year. We could alter any of these starting conditions without materially affecting the outcome. And, by the way, the sexual Regular rabbits can breed only with themselves, not with the Virgins.

We're off. The two Virgin rabbits and the two Regulars are released, nibble some grass, dig themselves into comfortable burrows and start breeding at once. The Regular male mates with the Regular female who, three months later, produces her litter of four rabbits, two males and two females. On the same day, though there was no sex involved, the first litter of Virgin rabbits is born. As both of the adult Virgins are female and each produces a litter of four baby rabbits, there are eight young born. So, even after the first litter, there are twice as many newborn Virgins as baby Regulars in the field: eight Virgins to four Regulars.

The new generation of Regular rabbits mate with each other and three months later, the two females each produce a new litter of four. There are eight rabbits altogether in these new litters, four males and four females. The same day the four Virgin rabbits, all being female, each give birth to four young. Now there are sixteen Virgin rabbits to eight Regulars. Here is where the disadvantage of sex begins to show. Four of eight Regular rabbits are male and can't get pregnant. So there are actually not twice but *four* times as many Virgin females as there are Regular females. With this arithmetic, after only five new generations, which is only eighteen months since we started, there are over a thousand Virgins in the field and only sixty-four Regulars. The meadow is slightly over-crowded but you get the point.

Even though I would be the first to agree that too much can be read into mathematical models, this very simple mental experiment has an obvious and clear conclusion: sex-free reproduction wins hands down. There are well over ten times as many Virgin rabbits as their sexual rivals. And the reason? The Regulars waste time and energy producing males that cannot give birth. All they do is to eat grass and inseminate the females. If we ran the mental experiment for a few more years in the same field and kept the rabbit population constant at around a thousand (maybe a family of foxes has come to live in the wood nearby and keeps the numbers down) the sexual Regular rabbits would soon become extinct.

Let us try one more mental experiment, this time more realistic. Suppose a lizard species which is reproducing in the normal sexual way finds a way of doing so without sex. This must have happened at some point to the ancestors of the whiptail. Let's suppose that the female lizards of this species usually lay a hundred eggs during their lifetime. Since the number of lizards remains more or less the same from one year to the next – because of the food supply, the predators and all the other things that regulate numbers in the wild – only two out of these hundred eggs will produce young that survive to breed – on average, one male and one female. Now imagine that a mutation happens in one of the females which allows her to reproduce without sex. How do she and her offspring perform compared to their contemporaries? Since she is no different from the other female lizards, except that she doesn't need sex to reproduce, only two of her offspring will survive. But they will both be females, clones who have inherited the ability to reproduce asexually. So, while the number of lizards overall has stayed the same, the number of asexual (or *parthenogenetic* – Greek for 'virgin birth') females has actually doubled – from one to two. This will happen with every new generation and the parthenogenetic lizards will quickly replace the normal lizards that still rely on sex. That is precisely what must have happened to the whiptail ancestors – they were driven to extinction by their asexual contemporaries.

These two mental experiments make it perfectly clear that sex is an extremely inefficient way to go about reproducing yourself, simply because the production of males is cripplingly wasteful. That being so, how is it that sex still survives as a method of reproduction in any species, let alone remains so very popular? The advantages of switching to sex-free reproduction, as the whiptail lizards have done, are so overwhelming that there has to be a very good reason for sexual reproduction to have persisted, in us or in any other species.

What started as a simple, almost childish enquiry has suddenly become a really difficult dilemma. We can now see that sex-free reproduction is far more efficient and that some species, like the whiptail lizard, have managed to break away from sexual reproduction to reap the benefits. But the curious thing is that if you look around for other examples of sex-free reproduction, they are found here and there in both the plant and animal kingdoms. However, with the important exception of the rotifers, where all five hundred species reproduce without sex, no large groupings of either animals or plants have completely abandoned sex, which is very surprising given all the advantages of doing so. The whiptail certainly has given it up, but it is very unusual in the lizard world. Not *all* reptiles, nor even *all* lizards have given up sex. Likewise, not *all* fish, nor *all* flowering plants reproduce asexually, though each grouping contains a few species that do. Could it be that other species which had managed to escape the inefficiencies of sexual reproduction have come to a sticky end?

Though changing your method of reproduction from sex to cloning is not physiologically trivial, plenty of animals and plants have managed to do it successfully, so the nuts and bolts of giving up sex cannot be all that hard to arrange – not compared to the huge theoretical advantages that life without sex has to offer, at least in terms of numerical success. But are the gains only short-term? Are there reasons why the apparent advantages offered by sex-free reproduction might only be relatively short-lived? The whiptail lizard is probably a relatively new species, perhaps only thousands rather than millions of years old. To put this another way: is the whiptail lizard in danger of extinction despite using a very efficient way of reproducing itself? Bringing the question closer to home, if we humans abandoned sexual reproduction in favour of cloning, would we be vulnerable to extinction? The crucial *genetic* difference between the two methods is that, in the sex-free species, the offspring all have exactly the same set of genes as their parents. They are all genetically identical – which is indeed what we mean by the term 'clone'. By contrast, the species that persists with the outrageously profligate process of sexual reproduction produces offspring that inherit a mixture of genes from both parents. They are all genetically slightly different. But why should that be helpful? This brings us right down to the basic process of evolution itself.

Though modified by nearly one hundred and fifty years of observation and argument, the principles of evolution by natural selection first systematically propounded by Charles Darwin in the mid-nineteenth century have remained solidly intact. How does evolution by natural selection help explain why genetically uniform species might do worse in the longer term than those with genetic variety? Darwin's theory, in a nutshell, is that the huge variety of different species in the natural world around us came about not through divine creation but by the slow adaptation of plants and animals to the changing environments in which they lived. For this to work, there have to be differences between individual members of a species which can be passed on to succeeding generations. Darwin was completely unaware of how this actually happened; he had no knowledge of genetics, chromosomes or DNA. He knew there must be a mechanism but had no idea what it was. As environments changed, Darwin argued, the individuals that suited the new environment better than their contemporaries tended, on average, to have more offspring, some at least of which would inherit the advantageous characteristic. The classic textbook example is the giraffe's neck: taller individuals are able to reach the highest branches, therefore having more to eat, therefore being better able to feed their offspring, thereby having more offspring survive than other giraffes, therefore increasing the neck length in succeeding generations. Looked at in this way, a species which is able to create more genetic variety, like new versions of the neck length genes, can evolve faster than a species which stays the same. Perhaps this is the clue we need to explain the advantage of sex over asexual cloning?

How does sex help create genetic variety? To understand this we need to bring in two basic aspects of genetics. One we have already encountered – the chromosomes; the other we have so far largely avoided. That is DNA itself. I will spare you a full account of DNA, what it is and how it works, and mention here only the briefest of details that are essential for our story. As we have seen already, DNA is, at its most basic, a long string of coded instructions, in many ways similar to a very long word. There are only four letters in the DNA alphabet, but even this restricted choice allows for, in effect, an unlimited number of combinations, or shorter words with different spellings. These DNA 'words' are the genes, which pass instructions to the cells to make the multitude of protein components which the body requires. As with any word in any written language, it is the sequence of letters in DNA that conveys meaning; change the sequence, even slightly, and you change the meaning. It is for this reason that the process of copying DNA, which must be done each time a cell divides, is so phenomenally accurate. If it were not, the instructions would very soon become irreversibly corrupted.

But, though the copying is extremely faithful, errors do very occasionally occur. These errors, called mutations – from the Latin *mutare*, to change – take different forms. At the simplest level, one DNA letter may be changed to another. Or a group of letters may get missed out. The cells, blindly obeying what they are told to do by the DNA, now follow the 'altered' instructions. In the majority of cases the changed spelling makes the word meaningless, leaving the cell unable to make a vital ingredient. Very rarely, a mutation will alter an instruction in such a way that it can still be read, and so the cell builds a slightly different version of whatever it is making. This could be a blood protein or an enzyme responsible for a particular bit of metabolism or a pigment with a different colour – or a bone protein that makes necks slightly longer. If these mutations happen in body cells, the effect might be felt locally – a patch of brown in an otherwise blue eye is a common enough example – but they will not be passed on to the next generation and do not count in the game of evolution.

Only if a mutation happens in a germline cell which goes on to produce eggs or sperm can it even audition for a part in the drama of evolution. It is then at least capable of influencing an individual in the next generation and maybe, just maybe, making it ever so slightly more successful, if not right away then in the future when the environment, in all its various guises, begins to change. It is hard for us to take in how anything so agonizingly slow could possibly have shaped so much of the living world; hard to believe that something so mechanical and yet so chancy lies so close to the heart of everything around us.

However, nothing I have said about DNA so far explains the advantage of sex. Mutations happen in the asexual whiptail lizards and they can be passed on just as they can in other lizards, or any other animal which has held onto its sex life. There is absolutely nothing in the mechanism of DNA mutation itself that favours sex over cloning. The advantages come not from the way in which mutations occur but in the way they are able to spread to other members of the species and, importantly, have a chance to join up with other mutations that they get along with. The key to this lies in the chromosomes and their odd behaviour just before eggs and sperm are formed, which we encountered in an earlier chapter. The two sets of chromosomes in the germline cells, one from each parent, have led separate lives, minding their own business and each completely oblivious of the other's existence. But shortly before they go off into their different germ cells they make contact, for the first and last time, and during their final embrace exchange long stretches of DNA. In scientific terms, they *recombine*. Since the stretches of DNA that chromosomes exchange are picked more or less at random, the recombined chromosomes are a unique combination of the DNA from both parents. Each egg and each sperm contains a new partnership of DNA never before seen in the history of the world.

This ability to create new combinations is inaccessible to the whiptail and all other species that have abandoned sex. A mutation in an individual sexless species – and mutations always happen in individuals – may well get passed on to her descendants and may well do them good and enable them to have more offspring. But it is always essentially working alone. Another mutation in a nearby gene on the same chromosome, if it occurs at all, will almost certainly happen in a different individual and it will only be inherited by *that* individual's direct descendants. It will never encounter the first mutation. Through the miracle of re-combination, only sex provides the opportunity for genes to mix. It is when the chromosomal exchanges which precede sperm and egg production bring together onto the same chromosome two favourable mutations, or introduce two mutations that alone are inconsequential but together are (or will be) dynamite, that any advantage of sex over cloning emerges for the first time.

This explanation for sex seems perfectly reasonable, and it is the one you will find in most school and college textbooks. Sexual species that shuffle their genes through recombination can evolve more quickly than asexual species which have forfeited their ability to mix and match. The cloners might get a short-term boost from cloning, but in the long run their rigid genetics cannot adapt as well to a changing environment. But things are not that simple – as we are about to find out.

9

THE IDEAL REPUBLIC

While the hunt for the elusive sex gene was going on in genetics laboratories lined with test tubes and furnished with microscopes, my own natural territory, a completely separate set of scientists were thinking about sex on a different plane altogether. These were, and are, the evolutionary biologists, evaluating and exemplifying Darwin's work on evolution and natural selection, then interweaving it with formal genetics into a comprehensive explanation of everything. Their laboratory is the natural world, its plants and its creatures from the forests of the Amazon to the deserts of Arabia – in fact, everywhere there is life. Their tools are observation and argument, equations and balance sheets. They see themselves, as one self-mockingly admitted, as the High Priests of the Ultimate Explanation.

Quite a number of them work, or have worked, in Oxford, as I do myself, but our paths only rarely cross and I enter their temple nervously. In preparing for this trespass, I found myself being constantly referred back to the scientific papers of one of their number, William Hamilton, who died only three years ago. Even though he worked in Oxford until his tragic and premature death, very few people outside the closed world he inhabited were aware of his brilliance while he was alive. Indeed, his modesty was such that I am told that even the fellows of his own Oxford college were quite unaware of the genius in their midst until they read his obituaries. But these left no room for doubt: '. . . the most influential biologist of his generation', '. . . one of the towering figures in modern biology', '. . . one of the foremost evolutionary theorists since Darwin'. I can imagine the bemused expressions on the faces of the other dons in the Senior Common Room as they read these words in the newspapers after lunch. The eulogies which run through his obituaries clearly mark him out as an exceptionally original scientist for whom the word 'genius' is not only a deserved but also an accurate description. But before he became one of the High Priests (and it was Hamilton who coined the phrase) he was, in his own words, a man tortured by self-doubt and given to long bouts of loneliness and despair. He describes in his autobiography how, as a research student at University College London, he was so utterly miserable in his bed-sitting room in Chiswick in west London that, rather than return home to his digs, he would spend hours after work at Waterloo station observing the hundred little dramas that are the essence of any great railway terminus.

Hamilton was, quite literally, a visionary. He was afflicted by what he called 'the migraine of evolution', a sense that almost within his grasp lay an explanation for the whole of nature, how it worked, and how it had come to be as it is. These visions were firmly rooted in the beguiling simplicity of Darwin's theory of evolution by natural selection. It is precisely this simplicity – that those individuals that are best at surviving and reproducing will pass on the quality that made them so to their offspring – which makes it so hard to believe that this principle alone is sufficient to account for all the complexity we see around us in the natural world. Not only does Darwin's theory need to convince us of its ability to maintain the extraordinarily abundant diversity of animals and plants which we see in the world and their, at times, fantastically sophisticated adaptations to a particular way of life, it must also explain how this all came about in the first place without the need for divine intervention.

Evolution by natural selection is an entirely automatic process, without any intrinsic morality. It is because of, rather than despite, its rational omnipotence that Darwinian evolution causes such deep unease among many people, such aggressive hostility among a few and such intoxicating enthusiasm among its devotees. It is such a very simple idea, yet it ensnares and beguiles with its tantalizing glimpses of the prospect of ultimate understanding. Hamilton himself was certainly intoxicated as he peeled away layer after layer of its secret disguises. But what strange and, at times, disturbing jewels Hamilton unearthed as he pursued his obsession.

His stunning scientific debut was to overturn the almost universally held belief that evolution works through the survival of the fittest individuals, as Darwin originally thought, or even for the 'good of the species'. Evolution by natural selection, Hamilton proved, works through *genes*. He first came to this profound conclusion, with its revolutionary consequences for contemporary biology, by solving the puzzle posed by that strange pattern of behaviour known as altruism. On the face of it, altruism – a deliberate act of self-sacrifice – goes against all the principles of Darwinian evolution. It is almost impossible to understand how individuals can promote their offspring's chances of survival, the essence of Darwinian evolution, by sacrificing themselves for others. How can it possibly help an individual to have as many offspring as possible by dying for another individual's – or the group's – benefit? The only circumstance in which it would seem to make any evolutionary sense at all is if you died in order to save your own children.

But altruism is not by any means restricted to humans; it is widespread among animals. Take just one example, from the hot plains of Africa. The meerkat, a small mammal related to the mongoose, lives there in colonies among a warren of burrows dug into the dry earth. While the others feed, one meerkat stands as lookout on a high point, perhaps a termite mound, so that he can warn other members of the colony of any approaching danger from snakes and other land-based predators. However, as he scans the ground from his vantage point, he is himself exposed to the danger of attack by an eagle from the air. He is literally risking his life to protect the colony. It is not hard to imagine how the colony benefits from his vigilance, but not at all easy to see what good it does the individual meerkat who exposes himself to such great danger while on guard duty. However, colonies that get too small to post a sentry soon die out. The natural interpretation is that this piece of altruism, even heroism, is being acted out 'for the good of the colony'.

The example of the meerkat and others like it gave rise to a school of evolutionary biology which explained the paradox of altruism along just those lines. Behaviours could evolve if they were for the common good even if they involved individual sacrifice. It felt comfortable to moderate the selfishness implicit in 'survival of the fittest' by diverting attention away from its emphasis on ferocious competition among individuals towards a mellower acknowledgement that actions which benefit others also have an intrinsic evolutionary value by contributing to the 'survival of the species'. This interpretation appeals to our ideas of charity and co-operation, which are so highly regarded. Group selection, as this modification of Darwinian theory came to be known, appeared to offer solid evolutionary support to political philosophies of socialism, even communism, where individual ambitions are subordinated to the benefit of the group, however defined – be it society or state.

Hamilton came to London as a research student already deeply suspicious of group selection. In the genetics department at Cambridge, where he was an undergraduate, the professor was the great geneticist R. A. Fisher. Along with the talented and eccentric English biologist J. B. S. Haldane, Fisher was one of only a handful of scientists who did not succumb to the temptations of group selection and stuck with the strict interpretation of evolution as working solely through individuals and the genes they pass on to their offspring. It is little wonder that Hamilton, exposed to this view as a student, was eager to solve the riddle of altruism without invoking group selection when he moved to London to start his PhD at the Galton Laboratory, part of University College.

The Galton Laboratory, and the professorship of its director, are named after Darwin's cousin Francis Galton, whose forays in the late nineteenth century into the inheritance of such characteristics as genius, feeble-mindedness and criminality (to use the contemporary terms) uncorked the bottle which left that stubbornly indelible stain on the history of genetics – the eugenics movement. This movement, which enthusiastically advocated selective breeding to enhance the genetic stock of our species, thrived in the years before the Second World War, particularly in the United States, the United Kingdom, Germany and Russia. Its ultimate disgrace was the bogus intellectual support it provided for the Nazi programmes of enforced sterilization and ultimately for the slaughter of those people considered genetically inferior.

Successive Galton professors have striven to erase the lingering odour associated with the eponymous chair, and this might explain the distinct lack of enthusiasm with which Lionel Penrose, the distinguished incumbent when Hamilton arrived in 1962, greeted the new graduate's proposal to explore the genetics of altruism. Hamilton's own version of Penrose's suspicions is characteristically colourful, as he records in the revealing autobiographical notes to his collected works, *The Narrow Roads of Gene Land*. 'Was I,' he wrote, 'a sinister new sucker budding from the roots of the recently felled tree of Fascism, a shoot that was once again so daring and absurd as to juxtapose words such as "gene" and "behaviour" into a single sentence?' Mind you, Hamilton was equally unenthusiastic about Penrose's work on genes and chromosomes, which he dismisses as 'elegant but rather mainstream'. Hamilton had already decided to ignore the revolution taking place in molecular biology since the discovery of the structure of DNA by Watson and Crick in 1953 had positively identified it as the ultimate embodiment of heredity; 'I was convinced that none of the DNA stuff was going to help me understand the puzzles raised by my reading of Fisher and Haldane,' he wrote. I am quite sure that several of the High Priests still feel equally condescending towards those whom they regard as mere artisans of the genome.

Hamilton's exploration of the genetics of altruism ranged widely over acts of apparent self-sacrifice throughout the animal kingdom which contradicted the basic principles of evolution by natural selection. In his view of the world, it was just plain crazy to think that any characteristic could ever evolve which was a disadvantage to the individual who possessed it. To Hamilton, that certainly encompassed any acts of self-sacrifice which resulted in the death of the individual concerned. How could that kind of behaviour ever have evolved in the first place? Yet the world is full of examples: bees that die after they sting; birds and animals that draw attention to themselves as they try to warn others of approaching danger from predators; even humans who risk their lives for their comrades in battle.

Hamilton's way round the paradox was to forget all about the fate of the individuals themselves and switch his attention to their genes instead. He began to think of the genes themselves, rather than the individuals that carry them, as the ultimate engines of natural selection. If, rather than concentrating on the survival of individuals – when altruism is almost impossible to explain – the focus is switched to the survival of genes instead, then the picture changes completely. In retrospect, it seems such a slight change of emphasis, but it was far more than that. It was a major piece of original thinking that has transformed biology over the past thirty years and created an entire new philosophy of biology, enthusiastically championed most notably by another Oxford biologist, Richard Dawkins. Hamilton realized that if natural selection acted on genes rather than on individuals or groups, then many pieces of otherwise contradictory behaviour could make sense.

The breakthrough came when Hamilton grasped that, although there is only one way for an individual to have offspring – by actually having them – there is another way for an individual's *genes* to get into the next generation. And that route is through relatives. If, by sacrificing his life on lookout duty, the meerkat saves the colony, his genes will still get to the next generation – not through him, but through his brothers and sisters. If an act of self-sacrifice increases the chances that his genes will survive through his relatives more than they suffer by his own death, then it makes perfect evolutionary sense – but only if genes rather than individuals are the route by which natural selection works.

This does take a bit of thinking about because it feels so counter-intuitive. It also begins to make us feel distinctly uneasy about ourselves. 'Are you telling me I don't count as an individual, that it is only my genes that matter?' I hear you ask. Well, in a sense, yes – that is the logical conclusion. But suspend your disbelief and your distaste for a moment and see where this path leads. Forgetting the X- and Y-chromosomes for the moment, we each have two sets of chromosomes, one from each of our parents. When we become parents ourselves, we pass on only one set of chromosomes, bearing half our genes, so that we share only a 50 per cent genetic identity with each of our children. However, we also share a 50 per cent genetic identity with each of our parents, and with each of our brothers and sisters.

Once Hamilton began to think of how genes might feel about this, he realized that they really don't care whether they get passed on to the next generation by you or by your siblings. They are happy so long as someone does it. So your genes are quite content for you to die if, by doing so, you save your brothers and sisters so they can pass on the genes instead. But how many of your siblings must you save by your heroism to meet with the approval of your own genes? Since each of your brothers and sisters shares 50 per cent of your genes, the basic algebra means that your self-sacrifice will be worthwhile if your altruism saves two or more of your siblings. Boiled down to its simplest ingredients, this means that, from your genes' point of view, it isn't worth dying to save one sibling and it's of pretty neutral benefit to die to save two siblings, but it's definitely worthwhile sacrificing yourself for three. It's sobering to realize that there are one and a half times as many of your own genes in three siblings as there are in you.

Of course, not all altruistic behaviour involves death, and Hamilton's treatment took this into account by quantifying the benefits of any behaviour pattern to the relatives and balancing those against its cost to the altruist. He also realized that, for the altruist, there were often more opportunities to help siblings or other close relatives, who are normally alive at the same time, than there are to help out grandchildren and further future generations. Nowhere is this behaviour more developed than in the bees and ants that live in large colonies where, to all appearances, individual effort is subsumed into the greater good. But what is really going on in these ideal republics?

Close to the library where I am writing this chapter stands Oxford's University Museum of Natural History. It is a master-piece of exuberant Victorian gothic architecture, completed in 1860. From the outside two storeys of well-proportioned windows along its broad frontage of creamy-yellow stone, reminiscent of the Doge's palace in Venice, rise up to support a high pitched roof pierced with yet further windows. I sometimes slip into the museum on my way to the library. Up the smooth worn stone steps and through the great wooden door, I come upon the great display hall, lit by a high glass ceiling supported on slender columns decorated in wrought ironwork. These days it is more devoted to dinosaurs and other children's treats than ever before, and today a newly acquired half-naked model of a woolly mammoth is receiving a fresh coating of red-brown hair. Around the hall, strong studded doors are topped by stone lintels confidently marked with the titles of the museum's original occupants – the Waynflete Professor of

Mineralogy, the Regius Professor of Medicine, the Professor of Geology, the Professor of Experimental Philosophy. These recall the museum's origins as the first real foothold of the sciences in Oxford, which until the mid nineteenth century was the more or less exclusive domain of the arts and the clergy. I pass slim columns of polished stone, each engraved with its own identity and origin – grey and white porphyritic granite from Lamorna Cove in Cornwall, pink granite from Peterhead in Scotland and dour grey and white granite from Aberdeen. Passing a fossil plesiosaur from Lyme Regis, I climb the broad yellow stone stairs leading to the gallery surrounding the great hall. Halfway up, as the stairway bends back on itself, is that laboratory of altruism – a beehive. The slight whiff of the camphor which, either real or imagined, perfuses the rest of the Museum is here sweetened by the gentle scent of old honey. The hive is cut away and, behind the glass, the colony pulses with activity in its daily routine.

Today is sunny and the workers are coming back from their foraging trips to the local flowerbeds, landing on the wide stone window ledge and crawling along the short wooden tunnel which leads to the hive. In among the huddle of bees one or two are performing their famous waggle dance, urgently vibrating from side to side then walking round in a circle and doing it again. This dance directs the other workers to their source of nectar by a complicated computation involving the waggle frequency, measured in waggles per second, the angle of waggle and the position of the sun. On the wall close to the hive a display of two concentric Perspex wheels with coloured arrows invites the onlooker, by turning the discs and factoring in the angle of the sun, to work out for himself where the flowers are from observing the dance of the bees. Try as I do, I can never get it right and sure enough, I fail again today. According to my navigation the bees are sipping nectar in the middle of St Giles – a broad tarmac road with not a flower in sight.

While some of the bees gather round the throbbing dancers, others are lodged headfirst in the hexagonal honeycombs, feeding the growing larvae. In one corner of the hive, five drones, larger and greyer than the workers, sit together as quiet and as immobile as old men on a park bench. Somewhere in the hive, though I can't see her today, the queen moves slowly from one cell of the honeycomb to another, surrounded by attendant workers as she lays a single egg in each one. She is the only bee in the hive to lay eggs. There are plenty of other females – the workers – but they are sterile and maintained in that childless state by hormones given off by the queen herself. The question that puzzled Hamilton was this. Why should every worker spend her entire life, all six weeks of it, looking after the queen and her children? It seems a complete waste of time and effort; they are not helping their own offspring, for they don't have any. Before Hamilton, the explanation was that all this self-sacrificing behaviour is for the good of the hive, and the good of the species – the classical reasoning of group selection. But Hamilton realized that the worker bees weren't doing all this to help the hive or the queen at all. They were actually doing it to help their own genes survive.

As we saw in chapter 7, bees do not use the XY method of sex determination familiar to us humans. Instead, they have a deliciously flexible system. The sex of a bee depends on how many sets of chromosomes it possesses. Females have two sets but males have only one because they develop from unfertilized eggs. The worker females have inherited one set from their mother, the queen, and the other set from their father, one of the drones. Though they live in the hive the drones don't lift a feeler to help with the housework, never go shopping for nectar and just sit around waiting for the chance to mate. How very different they are from human males.

The chromosome set which a worker gets from her mother has been scrambled by recombination so that each bee receives a slightly different mixture of genes. However, since the drones have only got a single chromosome set to pass on, all the workers having the same drone father will have received an identical set of genes from him. What does that do to the genetic relationship between the workers? It means that sisters are not 50 per cent identical, as in humans, but have 75 per cent of their genes in common. Hamilton saw that this is an even closer genetic relationship than a worker would have with any offspring she could have. Since she could give them only one of her two sets of chromosomes, a worker mother would share only half of her genes with her daughters, less than she already shares with her sisters. From the point of view of her genes, it pays to channel energy into helping the queen produce more and more sisters, with whom she shares a 75 per cent genetic identity, than it does to produce her own offspring, with whom the genetic identity would be only 50 per cent. The genes are very happy for the workers to remain sterile and keep on collecting the nectar. The altruism of the worker bee isn't for the good of the hive at all. It is for the good of her own genes – not in her, but in her sisters.

The demolition of group selection first begun by Hamilton threw the spotlight onto genes as the real engines of evolution. The lingering confusion over whether evolution worked through individuals or groups was cleared up. It was neither of these. It was the genes. Our genes are not serving us at all. It is the other way round. We are serving them – faceless, thoughtless, and ruthless.

This was even worse than straight Darwin for those people still looking for the hand of God in shaping the natural world. How could all this wonder have such a blind and mechanical foundation as the simple chemistry of DNA? This unsettling conclusion, deeply troubling to those who feel their fundamental beliefs have been shattered, has overshadowed all of biology for the past thirty years. There are those who disagree, certainly; but while the supremacy of the gene as the moving force in evolution has faced vigorous challenges from all quarters, it has not been overthrown. Now that, in the minds of biologists, genes had been released from servitude as the mere handmaidens of evolution into dominance as its principal agents, all sorts of things became possible. Genes could have their own ambitions. And in species like our own, where the two sexes have different genes, the possibility of separate motivations and the scope for struggles between them suddenly opened up. But there was one even deeper mystery to solve first.

10

THE SENSE OF SEX

After his triumph in overturning group selection and identifying genes themselves as the primary force in evolution, William Hamilton bent his agonizing mind to attacking an even greater puzzle – the evolution of sex itself. Could this perpetual enigma, so unlikely from the point of view of efficiency, as we have already seen, be solved by the same approach? Did sex and the multitude of inconveniences and waste it brings with it really evolve just to help genes? And, if so, which ones have most to gain and most to lose from sex?

The old-fashioned 'good of the species' explanation for sex was straightforward and comfortable. By enabling chromosomes to exchange DNA with each other, it provided a way in which a species could increase its genetic variety and so evolve more rapidly than its asexual, cloning counterpart. The sexual giraffe could grow a longer neck more quickly. The fatal weakness in the argument is that all the bother and waste of sexual reproduction is tolerated on the off-chance that the mixing of genes will benefit someone else at some uncertain time in the future. While it may be a good thing for the species to reproduce sexually rather than clone, because it does offer the chance of speeding up evolution, the burden of doing so falls on the thousands or millions of individuals who are going to get nothing for their trouble. Genes don't plan that far ahead. Conditions do not change all that much from year to year for most animals and plants to merit putting so much of their efforts into sex. Looked at like this, sex is beginning to sound just like altruism. It is a truly enormous burden which falls primarily on females who have to spend half their time and effort producing essentially useless males, their only satisfaction being that they may have helped the species evolve. Genes would certainly not approve of forward planning on such a grand scale. I can almost hear them. 'Not likely. We will be much better off in a female who gives up sex and puts all her efforts into producing exact copies of herself – and us. Let someone else worry about the future of the species.' It just doesn't wash. All the arguments are against sex, not for it. And yet it is all around us, the preferred method of reproduction for the majority of species – including our own.

It is a powerful argument, which has forced biologists to think of ways in which individuals and, more importantly, their genes could benefit from sex without appealing to any sense of altruism. After all, there are examples all around us of species that have given up sex and are doing extremely well, even adapting to very different conditions. Take dandelions. There are more than two thousand species of dandelion throughout the world, of which all but a handful have given up sex altogether. As any gardener knows, they are very successful plants and extremely difficult to get rid of. They produce golden yellow flowers in abundance and beautiful seed heads at which, as a child, I would blow to send the seeds floating off into the wind attached to their gossamer parachutes. Each of these seeds is an exact genetic replica of its parent – and of all the other dandelions in the neighbourhood. Nor can it be said that dandelions are restricted in their range. The same clones can be found as far apart and in such different surroundings as Greenland and Florida. They don't need sex for reproduction; nor, so it would seem, do they need sex to cope with changes in their environment.

Hamilton had the ultimate causes of sex on his mind when he

was working at the University of Michigan in the 1970s. During his walks to work he noticed that the woods were full of a familiar shrub, a species of blackthorn that brightens up the bare English hedgerows with its white flowers in the early spring. He knew it as the food plant of the bright yellow Brimstone butterfly from his boyhood in the Kent countryside. This wasn't a native American shrub and must have been imported from Europe. But, unlike the blackthorn bushes that grew in England, which were eroded by viruses and eaten away by caterpillars and other pests, the leaves of the Michigan shrubs were perfect and unblemished. The parasites and pathogens that plagued the blackthorn in its native Kent had not survived the transatlantic crossing. Here was a plant which, back in England, suffered continual attacks from its many tormentors. Was it using sex to outwit them? Hamilton began to imagine that the flowers of the imported blackthorn bushes were not really putting as much enthusiasm into their sexual role as the native plants – the milkweeds and the golden rods – were doing nearby. It was, he felt, almost as if they were considering giving up sex already.

Hamilton had glimpsed the one aspect of changing conditions that makes sex worthwhile. We are used to thinking of the environment in terms of the climate, the landscape, perhaps the supply of food – that sort of thing. But in evolution the environment is a much broader concept than that. It is all the other animals and plants, including the species you eat for food, and the species that eat you. These are the visible agents of evolution, the predators and the prey – the reasons for the swiftness of the gazelle, the camouflage of the tiger and the noiseless flight of the night owl. But what we do not see is the silent battle that goes on between all creatures and the unseen legions of pathogens that live on and beneath the surface of every individual. These are the bacteria, viruses, moulds and parasites that prey on all life from the simplest organisms to the President of the United States of America. We each think of ourselves as one individual, but in fact we are a combination of one very large individual and countless millions of smaller ones that live on and inside us. The constant, daily struggle between parasites and their unwilling hosts has shaped evolution more than any other force. And it is this perpetual battle which best explains why the majority of animals and plants are stuck with sexual reproduction.

A species which waves goodbye to sex makes itself extremely vulnerable to attack from pathogens and parasites. They too are evolving, changing their weaponry and their defences at every generation. And since they multiply much more quickly than their hosts, they can change very fast indeed if a new opportunity arises. Once a parasite has found a way of breaching its host's defences it can invade with alarming speed. If all the individuals in a species are genetically identical, as soon as a pathogen has unlocked the key in one, it can sweep through the entire species, killing them all. That is the very real danger faced by asexual species like the whiptail lizard and the dandelion. They risk the chance, almost the inevitability, that a pathogen will eventually pick its way through the succession of defences and, having done so, will wipe out the entire species in one devastating pandemic.

There are enough recent epidemics among humans and animals for us not to need reminding of the deadly speed with which a new pathogen can decimate a susceptible population: the Black Death in fourteenth-century Europe, which killed almost half the population; smallpox introduced from Europe, which killed millions of native Americans; and, of course, AIDS in Africa today. And that is in humans, a species with plenty of genetic variation among individuals thanks to the genetic shuffling organized through sex. A species that lacks this variation stands no chance. Had we been an asexual, cloning species with no genetic variety and susceptible to bubonic plague or smallpox, these epidemics would have killed everyone. In the sexual species there are at least a few individuals with genetic resistance which can survive the onslaught – though in the case of AIDS the irony is that sex is both destroyer and saviour.

This is, I agree, an extreme example of the dangers faced by the sex-free species – but the limited capacity they have to adapt to a changing environment, of whatever kind, is the reason they do not last for very long. If we could but see it, the fossil record is probably studded with extinct species that managed to shrug off the yoke of sex, and then paid the price. The long-term prospects for the whiptail lizard and the dandelion are not good.

This explanation for sex no longer smacks quite so much of unrealistic long-term planning. The individuals and the genes who pay the price of sex do not have to be quite so far-sighted. Locked in a constant struggle with your parasites, there is now all the more reason to produce offspring who are genetically different from you rather than the same, since your parasites will then have more of a problem feeding off your children than they did off you. Sadly, there is no promise of evolutionary progress here – you have to keep running to stay in the same place. And to do that fast enough you need sex. Viewed thus, sex is not quite such a piece of altruism, endured for the greater good, but an immediate way of outwitting the parasitic enemies that lie within you, and doing so in such a way as to enhance the chances of your own genes surviving to the next generation and beyond. Clones beware. You can run from your parasites but you can't hide. They will get you in the end.

In one final sad irony, it was a parasite that killed Hamilton himself. During an expedition to the Congo he caught malaria and, though he got back to England and seemed to be recovering, he was struck down by a cerebral haemorrhage which left him unconscious until he died five weeks later, in March 2000, aged sixty-three.

11

THE SEPARATION OF THE SEXES

Sex might be wasteful, it might be dangerous, but at least we now think we know why we do it – to stay one step ahead of the parasites that are always on our tail. There is one other vital aspect to sex which urgently needs an answer. It is the one that occupies us all of the time. Why are there two sexes? Why are there men and women? That's another of those questions to which, on one level, we all know the answer – you need two sexes to mate. But do you? If sex is just about shuffling genes around and exchanging a few with someone else, do they really have to be of a different sex? Can there be sex without sexes? Amazingly, there can.

To understand how, and why, this is so we need to travel way down the evolutionary scale, far beyond the animals and plants to more primitive organisms, microscopic creatures that do indeed have sex without different sexes. Some bacteria indulge in just that sort of sex, called *conjugation*. Two tiny bacterial cells line up alongside each other and, from the walls of one cell, a narrow pipe grows outwards until the two cells are connected. Along this pipe flow genes from one bacterium to its companion. When the transfer is complete, the pipe dissolves, and the bacteria separate and go their own way. The gene which forced this union and travels through the pipe has managed to spread itself around not only to succeeding generations, as when the bacteria divide in the usual way, but also to other contemporary bacteria. It is a long way from sex as we know it, but it does give a clue as to the origin of the whole strange process. Now we have realized who's boss in evolution, it is less of a surprise to discover that the gene that brings the bacteria together is also the one to sneak through the pipe. It has forced the bacteria into a primitive sexual liaison so as to be able to spread itself around. And any gene that manages to do that has a bright future. It is almost as if this gene achieves its ambition by infecting other cells after coaxing them into sex. Here we seem to have the inventor of sex and the first sexually transmitted disease rolled into one.

The bacteria that indulge in this primitive intercourse are not identifiably of different sexes. To see that distinction for the first time, to see the battle lines being drawn up, we must travel further up the evolutionary scale to microscopic organisms that are more complicated than bacteria but are still made of just a single cell. Unlike bacteria, whose single circle of DNA floats freely within the cell, these tiny creatures have chromosomes that are enclosed within a separate structure, the cell nucleus. Outside the nucleus, but still within the membrane which encloses the entire cell, is the liquid cytoplasm. This outer zone of the cell contains, among other things, the cell's protein assembly plant which obeys the instructions from the DNA of its nucleus. It also contains tiny structures called *organelles*: the mitochondria which contain within them the enzymes that cells need to use oxygen and, in plants, the chloroplasts which convert the sun's light to chemical energy.

The origins of the different parts of these simple organisms are still an enigma, but they are most probably descended from fusions of different kinds of free-living bacteria. The provenance of the cell nucleus is very obscure indeed, but the likelihood is that the ancestors of mitochondria were once bacteria that adapted to use oxygen. When the world was formed there was scarcely any oxygen in the atmosphere, and the gas we now think of as life-giving was first produced as a toxic waste. Tiny single-celled organisms called blue-green algae were the first to develop photosynthesis, the process which all plants use to harness the light of the sun, and oxygen is a waste product of photosynthesis. On a bright summer's day you can see tiny bubbles of gas slowly forming on the surface of pondweed and streaming up to the surface. This is oxygen. The molecules we take in with every breath started like this, as a by-product of photosynthesis in a plant somewhere. Chloroplasts, found within the cytoplasm of plants, evolved from the blue-green algae that had first discovered how to photosynthesize. Mitochondria evolved from bacteria that were able to turn the toxic oxygen waste from photosynthesis to their advantage by finding a way of using it to get more value from their food.

Even within these very modest single-celled organisms, built up from a nucleus and cytoplasm, there is already a confederation of different genomes from bacteria and algae working together in the same cell. Being animals and not plants, we don't have chloroplasts, but we certainly have a nucleus and mitochondria, the descendants of these bacterial ancestors, living side by side within our own cells. Each has its own set of genes, its own genome, and where there are different genomes there is the potential for conflict between them; as with any enduring co-operation between different parties, the benefits of working together have to outweigh the disadvantages. The independent and free-living ancestors of the mitochondria we all have in our own cells joined the confederation and took up residence in the cytoplasm. As bacteria, they had evolved the equipment to use oxygen and were the first to discover the intrinsic efficiency of aerobics. By switching from anaerobic to aerobic metabolism they managed to extract ten times as much energy from their food.

An ancient fusion between a mitochondrial ancestor and a cell containing a nucleus at once created a promising symbiotic situation. At least, there were big advantages for the cell that had hitherto been producing energy without oxygen. By involving mitochondria, it could immediately upgrade its energy production to use oxygen. It is not so clear to me what the mitochondrial ancestor had to gain from this arrangement. Maybe the bacteria that were the ancestors of the nucleus could do things that the mitochondrial ancestor could not and found useful. But perhaps it was more a case of the capture and enslavement of mitochondria by the cell with a nucleus than a mutually beneficial arrangement. How this fusion came about in the first place we will probably never know. Was the nuclear ancestor trying to eat the first mitochondria – or have sex with them?

Sex between these single-celled organisms is simply a matter of the fusion of one cell with another so that their DNA can be exchanged. That, of course, also happens in humans, when an egg joins with a sperm. But there are two very big differences. Single-cell sex involves cells of the same size, both containing a nucleus and cytoplasm and both looking very much the same. They are not male and female. So why have we needed to develop two sexes to achieve the same result? This was as great a puzzle to biologists as why sex evolved in the first place. Of course, we take it for granted, but it really isn't so easy to see why every species which has opted for sexual reproduction has evolved two sexes. In chapter 7 we saw that different species have come up with lots of different ways of creating the two sexes, from our own system based on the Y-chromosome to the probing hormone-laden tongue of the marine worm. But always there are two and only two sexes. Why?

Even though animals and plants are fantastically complicated conglomerations of billions of cells, each cell still retains the division between the nucleus and cytoplasm. In trying to find the answer to the question of the sexes, scientists had until recently rather overlooked the cytoplasm. Only the chromosomes in the nucleus have anything to gain from sex. That is where new gene mixtures appear, thanks to the recombination that accompanies sex. Everybody knew about the nuclear chromosomes; they were clearly visible under a microscope and their genetics had been well worked out, thanks in large part to the comparative ease with which their structures and their movements could be observed. Cytoplasm, on the other hand, was amorphous and the organelles within it were hard to see. As a consequence of this low visual profile, cytoplasm and the DNA it contains has been downplayed in its importance. There was a common presumption among evolutionary biologists that cytoplasmic genes coded for a small selection of comparatively trivial characteristics, a simple tune compared to the symphony of important genes carried on the nuclear chromosomes. But that attitude is changing fast. The cytoplasm, and particularly the mitochondria, are coming to be seen in their true significance.

Laurence Hurst, a PhD student of Hamilton's at Oxford and now at the University of Bath, and two American scientists, Leda Cosmides and John Tooby, became the champions of the hitherto disregarded cytoplasmic genes and have helped to promote mitochondria from their humble status as handmaidens of the nucleus to the major agent behind the creation of the two sexes. They were the first to see the genetic reasons behind that most enduring and unresolved conflict, of which we are all only too aware – the battle between the sexes. The two sides line up as follows. The nuclear genes, neatly assembled on their own chromosomes, see themselves as the all-powerful masters of the genome. They have worked out a way of relatively peaceful co-existence with one another which, save for the occasional outburst of revolt or dissent, has served them well. They may be ultimately concerned with their own survival and replication but they need to keep the vehicles in which they travel, the individual organisms (us, in other words), on the road. Running a whole organism requires the co-operation of many genes and, as we have seen, organisms need sex to keep one step ahead of their parasites. On the opposing side are the cytoplasmic genes. They have no need for sex, do not recombine, have never learned the meaning of peaceful co-existence and are, as we shall see, very good at pursuing their own interests. It was their violent objection to the fusion of cells that necessarily accompanies sex that was the crucial factor in the creation of separate sexes. To witness this titanic struggle you need go no further than the garden pond.

Put a drop of greenish water under a microscope and the chances are you will see a few tiny spheres like miniature emeralds gliding through the water. These are the single-celled algae called *Chlamydomonas*. The green colour comes from the chloroplasts, which capture the sunlight. The chloroplasts and mitochondria reside together in the cytoplasm of these tiny cells, separated from the nucleus by a thin membrane. All is peaceful as the tiny algae drift through the warm summer water. However, as soon as they start having sex towards the end of the season, all hell breaks loose. When the cells fuse as the preliminary to exchanging nuclear DNA, an immediate and outright battle breaks out in the cytoplasm. The organelles, the chloroplasts and the mitochondria from the two cells begin to slaughter each other with such brutality that only 5 per cent are left standing at the end. Their battlefield weapons are DNA-splitting enzymes which recognize and destroy the incoming cytoplasmic DNA. The nuclear genes can do nothing except stand back and witness this carnage as the organelles scratch and tear each other apart.

This war does no good at all to the cell, as the organelles' fight leaves a cytoplasmic battlefield strewn with the wreckage of combat. The nuclear chromosomes must respond to this destruction and in *Chlamydomonas* they limit the damage by weighting the outcome of the struggle so that the winner and the loser are already decided before the battle begins. To do that, genes on the nuclear chromosomes have created two different types of cell. Each is perfectly capable of living on its own, but one (the plus type) has more mitochondria and always wins the battle, while the other (the minus type), with fewer mitochondria, always loses. The nuclear genes, in their desire to fix the outcome, have arranged things in such a way by badging the outsides of the cells with identifying molecules, so that the only sexual fusions that can take place are between plus and minus cells. This way there is always a clear winner. Sex between cells of the same type, whether both plus or both minus, where both sides have evenly matched cytoplasms, would end in a draw, with no organelles left standing. These stalemates are prevented by the incompatibility of the surface molecules, put there by the nucleus, that dictate which cells can have sex together and which cannot.

Here at last we have the fundamental genetic reason why, in sexual animals and plants, there are two, and only two, different sexes. The separation of the sexes has arisen from a deliberate ploy by the nuclear genes to limit the damage caused by the two warring cytoplasms following the sexual fusions which the nuclear genes themselves require to exchange DNA. In a few organisms – including mushrooms, oddly enough – genes are exchanged by conjugation rather than by fusion. Narrow pipes connect the cells and only the nuclei are pushed through – the cytoplasms never meet. That is one way of avoiding the cytoplasmic war and it means there is no need to create two different types of mutually incompatible organisms, two different sexes. There is sex but not sexes. But our single-celled ancestors did not take that route. They decided to avoid the cytoplasmic war by creating the two sexes – and we are all living with the consequences of that ancient piece of diplomacy.

There is one other lesson we can learn from *Chlamydomonas*. Each of the tiny emerald spheres that drift through the water has only one set of chromosomes inside it. Their trigger for sex is when the nutrients in the pond, especially ammonia, begin to get low. Until then, while the going is good, they reproduce by simple fission, by splitting into two identical clones and carrying on. The reduction in nutrient levels is a signal that the pond which is their home is either about to dry out or, at least, is not going to be a suitable home for much longer. It signals that it is time to prepare for hard times ahead. So the sex begins: plus and minus cells merge and the fused cells, now containing two sets of nuclear chromosomes, develop a tough outer coating and prepare to sit out the hard conditions as spores. If the pond dries out, some of these spores may get blown by the wind to a new pool, like the rotifer tuns. Others stay behind in the ground, waiting for the rains to recreate their home puddle. When conditions do improve, the spores begin to germinate. First, the two sets of chromosomes inside the spore are doubled, then the whole cell divides twice and each of the four offspring cells is given one set of chromosomes. Finally, the tough skin dissolves and four tiny emerald spheres, each with one set of chromosomes, break out and glide off once more into their watery paradise to begin the cycle all over again.

Unlike us, *Chlamydomonas* spends most of its free and active life in sunlit pools with just one set of chromosomes. It spends the portion of the life cycle when it has two sets of chromosomes barricaded inside the reinforced walls of a spore, waiting for a release that may never come. That is completely different from us. Although we also have two distinct phases in our own life cycle, we, or our genes if you prefer, spend most of our time in diploid cells with two sets of chromosomes. I am writing this book, and you are reading it, with two chromosome sets in all of our body cells. The only haploid part of the human life cycle, when we exist with one set, is the time we spend as eggs or as sperm. It is hard for us to individualize these single cells and think of them as human, though they are all genetically unique. We might prefer to think of ourselves as advanced and complicated organisms. But our genes don't actually care where they are. They are as happy in sperm and eggs as they are in our body cells – if not happier. After all, they may have a future in eggs and sperm, which they certainly don't have in the cells of your body or mine, where their only prospect after a few years is to be buried underground or go up in smoke at the crematorium. While we like to think of ourselves on a completely different plane from pond life, we are only really thus in the stage of our lives that we spend with two sets of chromosomes. Then we can make a valid contrast between the human superbeings that we have become and the desiccated spores hidden in the mud of a dried-up pond. But during the portion of our life cycles that both spend with only one set of chromosomes, the human condition is remarkably similar to that of *Chlamydomonas*: floating around in liquid as single cells – but, in our case, without the aesthetic advantage of being an attractive green colour.

Now let us see just how the similarities play out. Just like *Chlamydomonas*, when our genes are preparing for sex, as eggs or sperm, they are inside two mutually incompatible types of cell. Egg does not fuse with egg, nor sperm with sperm. But the nuclear genes of our distant ancestors have taken the *Chlamydomonas* strategy a lot further. They have avoided the damage caused by the deadly cytoplasmic wars by entirely stripping the cells of one sex of its cytoplasm. We see here the logical conclusion of the strategy *Chlamydomonas* uses to fix the outcome of the cytoplasmic wars before they begin. What better way to avoid the conflict altogether than by denuding the single-cell stage, the *gamete*, of one sex of cytoplasm altogether? And this is exactly what has happened. The gametes of males, the equivalent of the handicapped minus-cell losers in the *Chlamydomonas* wars, have been systematically stripped of their cytoplasm until they are whittled down to a nucleus and very little else. These cells have become the sperm in animals and the pollen in plants. The single-cell stage of the female, on the other hand, has become the egg: a large cell absolutely bursting with cytoplasm and packed with literally thousands of mitochondria.

In masterminding the separation of the two sexes, our nuclear genes have stripped the male gametes of the wherewithal for an independent, free-living existence. The cytoplasm holds all the apparatus for the day-to-day workings of a cell, and no cell can survive for long without it. Sperm may have a nucleus with a full complement of chromosomes – but what use is that if there is nothing to carry out their orders? How could the downgrading of one gamete to a powerless gene bag with no capacity for prolonged independent existence possibly be achieved? The solution was to build up the other stage of the life cycle, the one with two chromosome sets, into a vehicle which is able to function in-dependently – at least until it can deliver the chronically disabled male gametes, deprived of their cytoplasm for the sake of peace, for fusion with the well-supplied eggs of the opposite sex. The encrusted spore of *Chlamydomonas*, sitting it out in the dried-up remains of a puddle, was thus eventually elevated from a temporary hideout for its genes during hard times to the most visible and active stage in the life cycle of most plants and animals – gamete delivery vehicles. Not only do we have to recognize the cytoplasmic peace deal brokered by the nuclear genes as the start-ing point for the separation of the sexes, we also have to thank it for the evolution of the fabulously complicated, multicellular, double-chromosomed organisms needed to protect and deliver the disabled and fragile gametes of the male sex. Men, in other words.

12

A WAR ON TWO FRONTS

The terms of the ancient peace treaty drawn up by the nucleus to halt the primaeval cytoplasmic wars had one fundamental flaw. In creating the two sexes, this treaty split every species into two camps and gave them opposing genetic interests – and we live with the consequences every day. We, like all other sexual species, are irreversibly segregated into male and female. Our own identity always begins with that definition. We are not first known as tall or short, kind or cruel, but as 'him' or 'her'. Our sex, our gender, is the preamble to any description of ourselves, and it fixes almost every aspect of our behaviour from the cradle to the grave. The creation of the two sexes might have ended an ancient war but it has replaced it with an enduring internecine struggle on whose battlefield we, like our ancestors before us, live out our entire lives, with men on one side and women on the other. But, unlike the straightforward savagery of the cytoplasmic battles of the past, where each side wanted only to destroy the other, ours is a more subtle struggle – a clash between male and female scripted by the characters of our own gametes and dictated by the terms of the old treaty drawn up by the nuclear genes.

The rules of engagement boil down to one simple fact. Under the settlement drawn up to end the cytoplasmic wars, one sex produces eggs, full of cytoplasm, and the other produces sperm, or pollen in plants, with a nucleus and not much else. No matter how sex is decided, it always ends up the same. Females make eggs and males make sperm. As we shall soon see, all manner of consequences spring from this one very simple distinction between males and females, between men and women. At this level, though, it is a struggle in which neither side desperately seeks outright victory. Each side needs the other. They might bruise and batter, but they don't want to kill. These sexual conflicts might be orchestrated by the nuclear genes, but they have no interest in the total victory of one side over the other. The nuclear genes are happy to watch the play, to observe the joys and sufferings of the cast, but they do not want the curtain to come down. Why should they? If one sex eliminated the other, the play would close and the audience of nuclear genes would have nowhere to go. Without a cloning back-up plan, they would immediately become extinct.

But the nuclear genes are not alone in our cells. They may want the sexual theatricals to run and run, but there is DNA in our cells that is screaming for it to stop. Cytoplasmic genes, such as mitochondrial DNA, are completely opposed to sexual repro-duction. Specifically, they have absolutely no interest in males. While the nuclear genes can afford to be pretty relaxed about the antics of the two sexes, since they get passed on equally well by either, mitochondrial DNA is not so even-handed. It has no interest in sex and gets nothing from it. Mitochondrial DNA does not experience the rapture of recombination, the shuffling of DNA which only nuclear genes enjoy. Quite the reverse: they are funda-mentally hostile to sexual reproduction, with the attendant futility of ending up in males half of the time. They pass from one gener-ation to the next only through eggs, not through sperm. A woman gives her cytoplasm with its mitochondrial DNA to all her children, but only her daughters will pass it on to the next gener-ation. Her sons convey it no further. Because sperm has been deliberately stripped down, practically to the bare nucleus, cytoplasmic genes are simply not transmitted by males. Sperm do hold a few mitochondria, just enough to provide the energy to work the tail, but at the moment of fertilization, when the sperm enters the egg, they are systematically hunted down and destroyed by a cytoplasmic defence mechanism dedicated to preserving the absolute supremacy of the egg's own mitochondria.

Barred from being passed down through sperm, cytoplasmic genes, which for us and other animals basically means mitochondrial DNA, have absolutely no interest in producing sons. Their own long-term future lies solely in future generations of daughters. Being in a son is a complete dead end for mito-chondrial DNA. Sex-free reproduction, with generation after generation of female clones, suits mitochondrial DNA very well. If mitochondrial DNA and other cytoplasmic genes are forced to endure sex and waste time having sons, can they do anything about it? They certainly can, as we shall see later. But if mitochondria hate sons, is there anything which loathes daughters just as fervently? A gene, or a piece of DNA, which has no interest in producing daughters for the same reason – that their long-term future lies elsewhere? And of course, there is. The Y-chromosome can get through to the next generation and beyond only through sons. Daughters do not have Y-chromosomes, so they do not count.

In the war zone that sex has created, there are two fronts. The first is where the perpetual skirmishes of male and female are acted out; where the strategies and tactics of the members of each sex ultimately depend on whether they are the ones to produce the eggs or the sperm, but where each is ultimately dependent on the other. The second is the site of the more sinister and more single-minded struggle in which two implacable genetic opponents, mitochondria and Y-chromosomes, fight it out. Each would happily eliminate the sex that did not serve its purpose – the sex on which the other depends to get it through to the next generation. And, as we shall see during the remainder of the book, they try very hard to do just that, and in ways that you never thought were genetic. We are getting very close to the essence of Adam's Curse: not one but two conflicting elements that give men and women different genetic agendas and mark out the eternal struggle between the two sexes whose consequences surround us every day.

13

A RAGE TO PERSUADE

The eggs of women are large, round, docile, self-sufficient, well provisioned with nutrients for themselves and their offspring, and produced in limited quantities – only one every four weeks. Sperm are the complete opposite. Stripped of their cytoplasm they are small, short-lived, frenetic and produced in enormous numbers – in men, at the rate of about 150 million every day. Women, their eggs outnumbered by sperm by several hundred million to one, can always be confident that they will get an opportunity to produce offspring. They are the guardians of a rare and precious thing – an egg. Men are not in this happy position. They must seek out and find a female willing to accept their sperm. Women can afford to be choosy, and since they keep the supply of eggs strictly limited, they have an interest in making sure that the sperm they allow to fertilize their eggs is from the best available source. In a great number of animal species, our own included, males spend an enormous amount of their time and energy persuading, or even duping, females into accepting their sperm rather than somebody else's. In fact, in a lot of species that is pretty much all they do. They are forced into becoming rivals and females choose between them. We have here the straightforward and familiar situation of supply and demand.

Impressing females is a costly business, as many of you know, and the lengths to which males of so many species will go to cajole females into accepting their sperm rather than a rival's is sometimes astonishing. But it obviously works, as the well-known example of the peacock's tail shows only too vividly. Growing and displaying the spectacular tail feathers with their iridescent greens, blues and burnished gold is a huge burden for the male peacock. It is cumbersome, heavy and dangerous, making the bird much more liable to being seen, and seized, by predators. But without a magnificent tail there is absolutely no chance of getting any sex. The drab yet presumably seductive females, safe in their camouflaged fatigues of brown and cream, demand and receive a full display before they consent to mate. If the show fails to impress then the peahen turns and retreats into the undergrowth, leaving the poor male disappointed and, literally, crestfallen. He packs away his finery and carries on with life until the next time.

The splendour of the peacock's tail is a direct result of what Darwin called *sexual selection*. As each new mutation arose to make the tail just that bit longer or the eye just that bit bluer it will have spread to succeeding generations through its ability to impress the females, who have also evolved a discerning eye that has a preference for such opulence. But what is it that the female really wants? The peacock isn't going to help raise the chicks and, after mating, the pair need never meet again. So quite why have generations after generations of peahens demanded to see the shimmering display? In a word, it's advertising. The peacock is signalling something else to the female – the quality of his genes. He is saying, in effect: I am so healthy and so strong that I can afford to waste all that energy on producing an intrinsically useless ornament – so my other genes must be absolutely sensational.

The world is full of other examples of sexual selection, where the preferences of one sex drive the evolution of features in the other which they find attractive in a mate. The supply and demand economics of sperm and egg production means that it is almost always the male who is trying to impress the female; the male who advertises and the female who chooses. As in any commercial campaign, only those males that do what the consumer wants reap the rewards. It is no good adding a new feature to the display that females don't appreciate. Peacocks may have beautiful tails but they can't sing. A peacock that could sing as sweetly as a nightingale would be wasting his time because peahens are not tuned in to song. Equally, a male nightingale with a brilliant blue-green tail would not make any impression on a female nightingale. Darwin realized that it was not just the features of the display itself that were evolving under the pressure of consumer choice, but the complementary ability to appreciate the product – and the desire for more of the same.

This was shown very nicely in an experiment with African widowbirds. The males have extremely long tail-feathers which they show off as they fly around their breeding territories. As you would expect, the males with the longest tails had most success in persuading females to mate with them. A team of biologists captured males and artificially shortened or lengthened their tails by cutting and grafting the central feathers with glue, then released them to see whether they did better or worse in attracting females than before the surgery. Sure enough, the males whose feathers had been artificially lengthened now did much better in that department, while the birds whose tails had been shortened suddenly found their seductive powers dramatically diminished. This straightforward test showed that the male's success in attracting females depended entirely on the length of his tail – not on his general vitality or on any other feature which the females on the ground were able to make out and factor in to their mating decisions. They were gauging these qualities indirectly by the extravagance of the tail. When the researchers released birds with tails surgically enhanced so that they were longer than any ever seen in the wild, these birds did best of all, irrespective of how puny their tails had been at the start. Clearly, the female widowbirds' appetite for longer and longer tails is still not satisfied, and the males will just have to try harder in the future.

There seems to be no consistency in which features are enhanced by sexual selection, and it may just have been chance which started the ball rolling in one particular direction. The ancestor of the first peacock probably just happened to grow a slightly showy tail, which just happened to appeal to a female. It could well have been something else – a slightly different head shape or a new way of walking. But once the male and female, advertiser and consumer, were on the same wavelength, they were both locked into an evolutionary spiral which exaggerated that particular feature and not others.

Darwin realized two things about sexual selection that set it apart from his earlier and better known theory of evolution by natural selection. The first was the speed with which it could change a species. Evolution by natural selection is usually excruciatingly slow, but sexual selection can transform a species extremely quickly, and where there has been rapid change, it is worth considering whether sexual rather than natural selection is at work. In this respect, no species has changed more rapidly than our own. Our immediate ancestors have conquered the world in less than a quarter of a million years since our beginnings in Africa. The common ancestor we share with chimpanzees, our closest primate relative, lived only six million years ago. These are long periods of time by our day-to-day reckoning, to be sure, but extremely brief in evolutionary terms. We certainly do have a lot in common with chimpanzees and other apes, but there are also a hell of a lot of differences: our upright posture, a very large brain, language, reasoning, art, superb manual dexterity – all features that are almost invisible in our primate cousins. All these features developed extremely rapidly in our ancestors, but failed to materialize in our close genetic relatives. Could this speedy transition have something to do with sexual selection? Did our male ancestors, with slightly larger brains and slightly better communication skills, slightly cleverer and slightly better with their hands, have the edge over their contemporaries, not so much in adapting to the external environment, but in getting more women to mate with them? Just as in the case of the peacock's tail, a successful campaign depends on a receptive and appreciative female audience who keep asking for more. But while the avian admirers could be appreciative without themselves growing gaudy plumage, our female ancestors would have needed to keep up, even keep one step ahead of the game. Eloquence is no use to a suitor whose object of desire doesn't speak a word.

The other feature of sexual selection that struck Darwin is that it can spiral out of control. The only check on the extravagance of the peacock's finery is not waning interest in showy tails on the part of the females but the incapacity of their suitors to do any better. As the experiment with African widowbirds showed, males could probably carry on growing longer and longer tails and reap the sexual rewards for their efforts until they could no longer lift off the ground. By the same token, will humans just keep getting cleverer and cleverer, or will their conversation become increasingly witty? If intelligence continues to attract women, then the answer should be yes. Then the question is: what are the limits to human intelligence or any other sexually selected feature? Will our brains get so big that our skulls explode? Unlikely.

In the example of the peacock's tail, sexual selection has added beauty to the world. That is, of course, only our opinion – so we must have a similar concept of beauty to the peahen's, though nowhere near as refined as hers when it comes to choosing between one shimmering tail and another. But sexual selection is not always productive of beauty; and not all females get to choose whom they mate with. Without doubt one of the least appealing mammals in the world is the male elephant seal. Unlike other marine mammals, seals must leave the water to breed. Good breeding beaches are few and far between, so, when they find one, elephant seals form large colonies of several hundred individuals. As on any beach, some spots are better than others. The best places are not too far from the sea (where the route to the water is blocked by other seals) and not too close to the sea either (killer whales have a nasty habit of launching themselves out of the surf to pluck a cub from the tideline). To seize control of the prime sites, male elephant seals have certainly not taken the route of evolving an elegant outfit or a melodious song. They have become revoltingly ugly, two-ton monsters. The reason the male has evolved these exaggerated proportions is to attract as many females as possible to his piece of beach and to keep them there, and, by fighting off intruding males, to ensure that he has more offspring than they do. This is ugly, brutish warfare and the males frequently inflict terrible wounds on each other with their vicious teeth, so that their furrowed necks are scarred and bleeding for weeks on end. Their life is so exhausting that the dominant males rarely manage more than one season at the top. However, for the winning male, the rewards for his genes are very impressive. In one study of a Californian elephant seal colony, 4 per cent of the males had over 80 per cent of the sex. Most males got no sex at all.

In contrast to the males, practically all the adult females did have sex and did have offspring, though a lot of pups ended up being crushed by the lumbering bulk of the beachmasters as they lunged across the prostrate bodies to ward off another amorous interloper. This is the epitome of the profligacy of sex. Vast amounts of food go to build up the enormous bulk of the males, which are often four times the size of the comparatively slim females. If that were not wasteful enough, the great majority of males never even get to mate and pass on their genes. Beaches are littered with the bodies of dead pups crushed by their mothers' jealous guardians. For anyone who still believes that evolution creates efficiency, look no further than the elephant seal. This is runaway sexual selection gone mad. No longer is the end result the extravagant yet essentially harmless display outfit of the peacock. This is bloody, brutal, murderous and ugly. Where will it end? Only when the male elephant seal grows so obscenely fat that he is unable to haul himself out of the sea onto the beach.

If sexual selection has orchestrated our own rapid progress over the last quarter of a million years, I wondered if I could read the signs of it in the genes we had inherited from our ancestors. The main genetic beneficiary of the successful bull elephant seal is his Y-chromosome, passed on to all his sons at the expense of those of the seals that could only stand and watch. Could I find signals among our own genes which revealed a history of sexual selection in our own species, whether orchestrated through the same gentle persuasion as in the peacock and the nightingale or by the oppressive brutality of the elephant seal?

14

MEN OF THE WORLD

It is a long time since, tossing pebbles into a Yorkshire stream, we contemplated the Sykes Y-chromosome. Since then we have tunnelled through very hard soil to excavate the basic mechanisms of sex, the fundamental reasons for it, the rationale behind the creation of two sexes and the power of sexual selection. To be perfectly honest, I never imagined, when I grabbed the DNA brush to swab Sir Richard's cheek, that I would become so deeply embroiled in such utterly fundamental processes. I had spent many enjoyable years collecting and interpreting mitochondrial DNA from volunteers all over the world and using it to unravel bits and pieces of human history. It was natural enough for me to be interested in the Y-chromosome, but, as I explained right at the beginning of this book, I saw it only as another way of doing much the same thing; a means to reconstruct a story about our ancestors and our own evolution from the DNA that had percolated through to the present day. I don't think anyone who was involved in that sort of molecular prehistory paid any real attention to the fundamentally different characters of the two pieces of DNA that they were reading. I think we were all too preoccupied with our own favourite to concern ourselves with these much deeper issues.

In fact, it was even sadder than that. I regularly attended seminars by Y-chromosome *aficionados* at which mitochondrial DNA was criticized for being far worse at unveiling the human past than their own preferred segment. I was used to listening to complaints that mitochondrial DNA was only a single genetic system among many, and as such capable of giving only a limited picture of human history. The fact that mitochondrial DNA was a single system was undeniable, but I remember thinking at the time that this was a mean-spirited accusation. It sounded rather like criticizing the Apollo programme for landing men only on the moon rather than on all the planets at once. The absurdity of the situation peaked, for me, at a lecture given by a leading geneticist who began his presentation with a withering denouncement of mitochondrial DNA for being just a single system – and then went on to extol the virtues of the Y-chromosome, on which he was working. It was one of those pathetic 'my gene is better than yours' situations which I honestly find quite pitiable. Both genetic systems are valuable and both have outstanding abilities to trace the separate history of men and women – and, when taken together, to probe how the two sexes have interacted in the past.

Over the course of the last fifteen years, DNA has helped to unveil many secrets of our evolution hidden by millennia of myth and legend, plotting the history of our species from its earliest beginnings in Africa to its present domination of the planet. This has not been the triumph of genetics alone, but has been achieved by fitting and blending the new and independent DNA evidence with the existing disciplines of archaeology, palaeontology and linguistics. It was mitochondrial DNA which ignited the genetic revolution, and the Y-chromosome which has consolidated it. Their dual success is due to their total commitment to one sex or the other. Because each is inherited from only one parent, their histories are not muddled by recombination like the rest of the nuclear chromosomes and they are excluded from the exchange of DNA between chromosomes that is the reason for sex. The ancestral echoes reverberating from the past are very much simpler to interpret from the DNA of mitochondria and the Y-chromosome than from the ever-changing nuclear chromosomes. Nor is recombination the only issue. The genes on your nuclear chromosomes (other than the Y, if you have one) have been inherited from both your mother and your father who, in turn, have inherited theirs from your four grandparents. Even over just two generations it is practically impossible to say which gene has come from which grandparent. Twenty generations ago, one of your nuclear genes could have come from any one of a million different ancestors. The trail is just too confusing to be useful. The direct and unscrambled links to the past followed by mitochondria and Y-chromosomes are mercifully untroubled by these complications. However, although they share a parallel inheritance, the detail of their DNA is quite different.

To begin with the simpler of the two, every one of the hundreds or thousands of mitochondria within each cell has its own DNA. This is formed into a small circle – just as in the bacteria which were its ancestors. Compared to the enormously long linear DNA in human nuclear chromosomes, our mitochondrial DNA circle is minute. It holds only 16,569 bases (the 'letters' of the DNA code), to be absolutely precise. Still, though small, it is packed full of genes: thirty-seven of them altogether. In our nuclear chromosomes, genes are separated or even interrupted by long stretches of meaningless junk, but the mitochondrial genome is a masterpiece of precision and frugality. No waste DNA separates the genes and they are arranged, one after the other, head to tail around the DNA circle with no space in between. Only one stretch of mitochondrial DNA, called the *control region* and roughly a thousand bases long, contains no genes – but it does have two very important purposes. First, the control region is where the circle starts to copy its own DNA when it needs to divide; and it is also the starting point on the DNA circle for the genes to be read, rather like twelve on a clock face where one circuit ends and another begins. The good news as far as I was concerned is that these vital functions of the control region can be carried out without its having to consist of a particular, precise sequence of DNA. Almost any sequence will do so long as it is a thousand bases long. The reason this is good news is that it means mutations, when they occur, can do no harm and so are not eliminated by selection. Because of this they accumulate in the control region much more quickly than elsewhere in the mitochondrial DNA circle. To read the messages from the past I needed DNA with plenty of variation between individuals, where mutations had built up over time and were still there to be interpreted.

Mutations in DNA occur only exceptionally rarely as it is copied. The most common mistake is when one base simply changes to another one – an *A* to a *G* or a *C* to a *T*, for instance. Since DNA is nothing but a set of coded instructions, whether a mutation like this will have any effect depends on which in-struction is changed and how it is changed. If the mutation changes a letter in a vital part of a gene and sends out a completely different instruction, when it is being read the mitochondria, which follow these altered instructions completely blindly, will then make a different version of whatever the gene is controlling, which is usually the details of how to make a particular protein. Proteins made with their instructions altered by mutation will almost always work less well than the original. Natural selection will sooner or later eliminate the individuals who carry these mutations and they will disappear from the face of the earth. Such harmful changes as these were no use to me if I wanted to use the DNA of living people to explore the past. They will have disappeared long ago.

That is why stretches of DNA that are not genes, in the sense of carrying instructions, are so very useful. When rare random mutations strike these regions the sequence of DNA will change, but that doesn't really matter. Since there are no instructions, none can be changed. The mitochondrial control region is perfect in this respect. Mutations can occur to change a DNA base more or less anywhere within the control region but, because the precise sequence of bases is unimportant, it will not make any difference to the mitochondria which carry it. It will not have any effect at all, harmful or otherwise. Individuals who carry mutations in their mitochondrial control regions will be neither eliminated nor encouraged by natural selection. The mutations can survive in the descendants of the people inside whom they happened and remain there for us to find and interpret.

Over the last ten years or so, my colleagues and I have read through the control region DNA sequences of thousands of individuals from all over the world. These sequences are a fabulous treasure trove of variation – the lifeblood of genetics. Even now, almost every week I come across a new sequence that I have not seen before. As there is so much variety, it means that when we do find two people who have exactly the same sequence in their mitochondrial DNA control region, they are almost certainly both recently descended from a common ancestor. I say 'almost certainly' because mutations can happen twice in the same place, but this is rare. Since mitochondrial DNA, or mDNA for short, is maternally inherited, two people with the same control region sequence don't share just any old ancestor. It has to be one to whom they are both connected by an unbroken maternal genealogy.

As we built up mDNA results from more and more people, my colleagues and I began to see that they fell into broad groups. We saw large numbers of people whose mDNA was not absolutely identical but was sufficiently similar to indicate a shared ancestry somewhere in the past. Among Europeans we discovered seven such groups, whose individual members were all related through their mDNA. In other parts of the world we found different groups – thirteen in Africa, four among native Americans, and a further eleven in Asia. By logical deduction each of these related groups, which I called clans, must have at its centre the DNA of just one woman from whom all the members of the clan are maternally descended. She is literally the maternal ancestor of them all, the ancient mother of the entire clan.

Since I knew the rate at which mDNA mutations occurred, I could work out approximately how long ago each of the clan mothers had lived. I did this by adding up the mutations that had accumulated in her various descendants whose DNA we had studied and then multiplying that figure by the mutation rate, the rate at which DNA changes with time. The detail of this process is recounted in my earlier book, *The Seven Daughters of Eve*, and I won't repeat it here. Not surprisingly, these women lived a very long time ago and each has tens or hundreds of millions of maternal descendants living today. The seven European clan mothers, according to these estimates, lived at different times between something like 45,000 years ago for the oldest and only about 10,000 years ago for the youngest. Six out of the seven clan mothers lived well before agriculture spread into Europe and I used this finding to argue that the maternal ancestors of most modern Europeans had been hunter-gatherers rather than farmers, as had previously been widely thought. Finally, we were able to connect the different clan mothers from all over the world through their own ancestors and draw out an enormous maternal family tree for our entire species. The maternal connections brought to light by mDNA were useful in another context as well. They enabled me to trace the movements of our maternal ancestors across the globe by linking together their present-day descendants. I was able, for example, to show the genetic connections linking the original settlers of the far-flung islands of the Pacific with southeast Asia rather than with the Americas. Amid the enthusiasm, though, I tended to forget what history it was that I was actually reading. It wasn't a history of our species, only a history of women.

Compared to the small and compact circle of mDNA, the Y-chromosome is enormous. From one end to the other of a typical

Y-chromosome, there are about sixty million DNA bases rather than the sixteen and a half thousand in the mitochondrial circle. Research on the Y-chromosome ran a few years behind mitochondria for what was at first a practical reason – all Y-chromosomes seemed to be exactly the same. This was a surprise because everyone expected Y-chromosomes to come in a multitude of genetic varieties, just like mDNA. Such optimism was founded on the enormous amounts of junk DNA that nuclear chromosomes were known to contain. Just as the precise sequence of the control region in mDNA is unimportant, so the sequence of junk DNA doesn't seem to matter – so any changes brought about by mutation will not alter important instructions. Cells take no notice of the sequence of junk DNA, so these changes will not be scrutinized and eliminated by natural selection and will be passed on to descendants. That is why everyone expected Y-chromosomes to be just as rich in variety as mDNA obviously was. When that turned out not to be the case there was a worrying hiatus for a short while. However, there were differences to be found and a handful of devoted laboratories, after removing tons of amorphous topsoil, unearthed a few rare jewels. These were single-base changes in the DNA of Y-chromosomes, just like we had found earlier in mDNA. Predictably, these few changes occurred not within genes but in the DNA between genes and whose precise sequence escapes the spyglass of selection. After many years of late nights in the lab and sequence after sequence of frustrating uniformity, enough differences were found to enable researchers to begin sorting out the spectrum of Y-chromosome variation.

The existence of identical Y-chromosomes hinted at a common paternal ancestry, in an exact analogy to the maternal connections revealed by mitochondrial DNA. Just as mDNA sequences group people together through their links to a shared maternal ancestor, so Y-chromosomes began to reveal groups of men related through their fathers. By a mirror-image logic to that of mDNA, but this time tracing the paternal lines instead of the maternal, men in the same group have to be descended from just one man – their clan father if you like. Little by little, these rare differences were used to draw up a paternal family tree of our species equivalent to the maternal tree laid out in *The Seven Daughters of Eve*. It is still quite sparse, but I nonetheless reproduce a recent version in figure 3, drawn out from 153 different Y-chromosomes from men living in many different parts of the world. Just like the maternal tree of our species, this is an interconnected network of genetic clans, each represented by a circle. Each of the fifteen circles represents a cluster of Y-chromosome genetic fingerprints from several men which, though they are not necessarily identical, have a lot in common. Men in the same clan usually come from the same continent and they have genetic fingerprints which share features in common, and the inescapable deduction is that they are descended from a common paternal ancestor. It is still a very basic tree and will improve a lot over the coming years as more and more men have their Y-chromosomes analysed.

The connections the tree makes are deep ones, going back ultimately to one single paternal ancestor of all males by the same inescapable logic which sees us all maternally descended from one woman. The logic is the same but the connections are different. The links to Y-chromosome Adam are through the paternal genealogy traced out by the Y-chromosome. Our connections to Eve are through the maternal line, revealed to us by mitochondrial DNA. There are many differences between the mitochondrial tree and the early tracings of the new Y-chromosome family trees, but they are not fundamental ones. There is no doubt at all that both have their origins in Africa. That is where the deepest roots of humanity lie and this is shown very clearly on both trees. All the earliest branches of both trees strike off from the trunk in Africa. There is no doubt that, just as mitochondrial Eve lived in Africa, so did Y-chromosome Adam. But Adam was not the only man around at the time, any more than Eve was the only woman; nor did they live at the same time. They just happen to be the only individuals whose

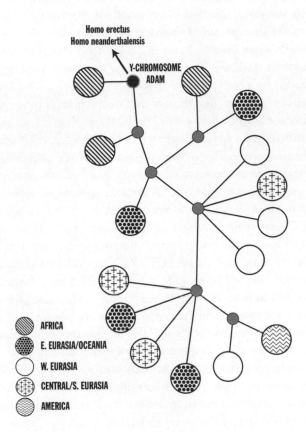

Figure 3: Men: the world clans

paternal (Adam) or maternal (Eve) genealogies stretch down unbroken to the present day. The paternal lineages of descent from Adam's contemporaries did not make it, because they ended either in a childless man or in a man who had only daughters. Likewise with Eve: only her direct maternal descendants have survived to the present day. The maternal lines of the other women alive at the same time as Eve came to a dead end either because they had no children or because they had only sons and not daughters.

In retrospect, I suppose there were two surprises in the way the Y-chromosome work had gone. The first was that it had been far more difficult than anyone had predicted to find genetic variations in Y-chromosomes. The second was that the age estimates from the Y-chromosome tree showed a much more recent ancestry for *Homo sapiens* than did the equivalent estimates from mitochondrial DNA. Eve, it appeared, lived a lot longer ago than Adam. These figures, based in both cases by adding up the mutations and multiplying by the mutation rate, had Eve living 140,000 years ago and Adam only 59,000 years in the past. At the time, in the late 1990s, nobody took a lot of notice. Labs which had sweated to find Y-chromosome variants were just glad to get the work over with. The discrepancy between the ages of Adam and Eve was put down to inaccurate assumptions about the mutation rate or something equally mundane. Only much later did I realize what this discrepancy actually meant.

I was keen to apply the new results from the Y-chromosome work in a place I had got to know very well through my work on mDNA – the beautiful Polynesian island of Rarotonga, in the Cook Islands, five hundred miles east of Tahiti. I had been entranced by these islands since I had first visited them in 1990. Sprinkled on the blue vastness of the Pacific like a thousand tiny emeralds, the islands of Polynesia have captivated Europeans since Magellan made his first voyage across the ocean in 1520. But he was not the first to arrive. Over two thousand years earlier, the original

Polynesians had turned their voyaging canoes into the wind and set out on the most audacious maritime explorations ever undertaken by our species. With no prior knowledge of what lay ahead, they had crossed the hundreds and thousands of sea miles which separated these scattered specks of land. By the time the Europeans began to arrive in the sixteenth century, the Polynesians had already reached every single island in the Pacific. Some had since been abandoned but most, from Hawaii in the north to Rapanui (Easter Island) in the east and Aotearoa (New Zealand) in the far south, were permanently settled by the ancestors of these most accomplished of ancient seafarers.

It was while staying on Rarotonga that I had first realized the power of mDNA to unravel human history. From a few blood samples from Rarotongans taken from the local hospital, which I attended after breaking my shoulder in a motorcycle accident, and later from a much wider survey, I discovered that the mDNA of most Rarotongans was practically identical and that all had its origins in south-east Asia. This was a blow for the supporters of the late Thor Heyerdahl's theory which had the Polynesians sailing to these islands from the shores of South America. But this was mDNA and I was reading the history of women. It was still formally possible that the men had come from South America and the women from south-east Asia. Possible, but ludicrously unlikely. Nonetheless, as soon as I heard about the new Y-chromosome markers, I wanted to clinch the question by showing that both the men and the women had come to Polynesia from the same direction.

From my original collection, I picked out the DNA samples from thirty-three Rarotongan men. I knew already, from my earlier research, that these men had mDNA sequences which were identical to one another. I also knew from these sequences that these men were, without any doubt, maternally descended from the original Polynesian settlers. They all had a characteristic sequence that was unmistakably Polynesian. Through their mothers, they had a clear and direct connection to the women who had travelled on board the first canoes to head out into the unknown waters of the Pacific three thousand years ago.

But had their Y-chromosomes taken the same route to the island? In order to answer that question I joined forces with geneticists Mark Jobling and Matt Hurles from the University of Leicester in central England. Mark's team were part of a consortium of laboratories in Europe who had been developing, over several years of patient research, one of the early genetic marker systems used to distinguish the different clusters of Y-chromosomes. Mark's own system divided Y-chromosomes from around the world into about two dozen different clusters based on a series of marker mutations. The Leicester system, which contributed to the tree shown in figure 3, supplied a broad framework for human evolution seen through the history of men and, as with mitochondria, the different clusters could be interconnected to show how one had evolved from another. The mutations which distinguished one cluster from another occurred so infrequently that it was reasonable to assume that they had happened just once during the whole course of human evolution. If two Y-chromosomes fell into different clusters, that meant they could not be closely related.

While Matt was working through the Rarotongan samples on the basis of Mark's system, Jayne Nicholson, who had recently joined my research team as a graduate student, set to work on analysing the samples for a different marker system altogether, one that I would also use to trace the Sykes Y-chromosomes. Compared to the slow genetic changes being mapped out in Leicester, Jayne's system was fast and furious. The Leicester system followed the single mutational changes in DNA that happened once in a blue moon and divided up Y-chromosomes into a few separate clusters. That was fine as far as it went, but it meant we could distinguish only broad classes of Y-chromosome. If we wanted any finer detail then we had to find a way of telling the difference between Y-chromosomes *within* each of the clusters. To do this we needed to work up an entirely different marker system from the Leicester group, one that was deliberately chosen for a fast rather than a slow mutation rate. Eventually we settled on the same sort of mutations that are used by forensic laboratories for genetic fingerprinting. These are not the very rare DNA changes that had happened only once during human evolution and which had proved to be so valuable to Mark for dividing Y-chromosomes into separate clusters. The markers we chose relied on a different type of DNA change altogether.

Some DNA sequences, when they crop up in the human genome, seem to take on a life of their own. Rather than being exactly the same for generation after generation, save for the very occasional mutation of the type we have already encountered, these pieces of DNA change very quickly, even between one generation and the next. These volatile segments are short stretches of DNA which are repeated over and over again – ten, twenty, fifty, even a hundred times. The actual sequence that is repeated is commonly two, three, four or five DNA units, or bases, in length. When DNA comes to be copied, the normally extremely reliable mechanisms for ensuring pinpoint accuracy in the copies don't seem to be able to cope with these rogue repeats anywhere near as well as usual. The repeats somehow seem to jam the normally unflappable molecular inspection system that detects and corrects DNA copying errors. What starts out as, say, ten repeats of a short sequence like *CAG* can quite easily become nine or eleven repeats in the copy. The actual sequence itself is accurately reproduced but the number of times it is repeated is not. It can go up or down, usually by just one repeat, as in the example, but sometimes by two and, very occasionally, by three in one jump. Every time one of these length changes occurs it counts as a mutation.

There are hundreds, perhaps thousands of these unstable repeats scattered on all nuclear chromosomes, and the Y-chromosome is no exception. They take a bit of finding, but once you track one down it can become a gold mine of variation. The different numbers of repeats can be told apart by measuring the overall length of the segment, which is easy to do. It is simply a case of forcing the repeated segment through a gel with an electric current and measuring its speed with a laser detector. Short segments, with only a few repeats, move more quickly through the gel than longer segments with more repeats in them. In the example we used just now, a segment containing the three-base sequence *CAG* repeated ten times will be thirty-three bases long, three bases longer than a segment containing the same sequence repeated only nine times and three bases shorter than a segment with eleven repeats. It isn't much, but modern equipment is easily capable of telling the difference between the two.

The beauty of these markers for our research was that we could often distinguish not just two different Y-chromosomes but three, four, five or even six different Y-chromosomes with the same marker. That's because there are lots of possibilities for the total number of repeats that a segment may contain. A *CAG* sequence repeated ten times on one chromosome might show eight, nine, eleven, twelve or thirteen repeats in other Y-chromosomes. That is great in itself; but the real benefit comes when we combine the results from more than one marker. Two different markers at different places on the Y-chromosome, each with, say, six different numbers of repeats, gives $6 \times 6 = 36$ possible combinations. Include a third marker, also with six possibilities, and the combinations increase again to $6 \times 36 = 216$ different Y-chromosomes that can be told apart. By the time ten markers, each with the same amount of variation, are combined, the number of possible combinations goes up to 6^{10}, which works out at 60 million. It is that kind of algebra, combining information from several different markers, which makes forensic genetic fingerprinting so very good at individual identification. However, we were not doing standard genetic fingerprinting, which involves working with markers on the other nuclear chromosomes; we were creating our own equivalent system for the Y-chromosome alone.

What appealed to me was the possibility of combining these two marker systems. On the one hand, Matt and Mark could distinguish two dozen or so different clusters of Y-chromosomes. On the other, we could tell literally thousands of individual chromosomes apart using our fingerprinting method. If we could do that, you might well ask, why did we need the Leicester system at all? The reason is that our markers, by their very nature, change very quickly. Since in Polynesia we were looking at events which happened over thousands of years, our Y-chromosome fingerprints might have changed so much in that time that we would not know whether two Y-chromosomes with very different fingerprints were related back to a common ancestor or not. If we suspected a common ancestry, it would be reassuring to know that the Y-chromosomes were in the same cluster defined by the Leicester system based on slow change. Likewise, there was a real chance that identical, or near-identical, fingerprints in our system might mislead us into thinking the two Y-chromosomes were closely related when, in fact, they were not. Confirmation that they were in the same cluster, as defined by Leicester, would tip the balance in favour of a close relationship. Y-chromosomes in different clusters could not be closely related even if their finger-prints appeared to suggest it.

When we sat down to compare the results of our joint endeavours, the results were intriguing, to say the least. Even though we had picked thirty-three Rarotongans on the basis of their having exactly the same mDNA, Matt had found that their Y-chromosomes belonged to three separate clusters. For convenience I'll refer to them as A, B and C. Right away we had discovered that Polynesian Y-chromosomes, in Rarotonga at least, were more varied than Polynesian mDNA. In Rarotonga, by far the common-est of the clusters was A, embracing nineteen of the thirty-three chromosomes. When Jayne showed us her fingerprint results on these nineteen men, it was immediately obvious they were all very closely related to one another. This was so reminiscent of the mDNA findings from Rarotonga that we immediately assumed, correctly as it turned out, that these were the descendants of the original Polynesians who had settled the islands from the west. We proved this a couple of years later in a more comprehensive survey of other Pacific islands along the route, when we found cluster A-chromosomes scattered all along the route. These nineteen Rarotongans could count the very first settlers of this remote island as their maternal *and* their paternal ancestors. The mDNA which fuelled their bodies and the Y-chromosome which made them men had been brought across the thousands of miles of open ocean on board the first canoes.

Though there were fewer of them, the four Rarotongan Y-chromosomes from cluster B also had closely related fingerprints, and we took this to be another set of chromosomes that had been passed down from the first arrivals. We had found a few among Papuans living on the coast of New Guinea and, later, on other Indonesian islands. That increased the number of Rarotongan men who could claim both maternal and paternal descent from the first settlers from nineteen to twenty-three out of the total of thirty-three. But what of the other ten? Had these arrived from America? We had not yet seen any evidence of native American Y-chromosomes. Were they waiting to be discovered among the ten remaining chromosomes in cluster C?

Could the cluster-C Y-chromosome fingerprints give us any clues? Jayne called them up from her computer, and – to our surprise, and in complete contrast to the other Polynesian chromosomes – every fingerprint was different. They were in the same cluster as defined by Mark, to be sure, but these Y-chromosomes were not closely related in the way that the Rarotongan chromosomes from clusters A and B had been. That made it very unlikely they had arrived on Rarotonga with the very first settlers from Indonesia and south-east Asia. The succession of islands that marked the route of the original Polynesians – New Guinea, New Britain, the Solomons, Santa Cruz, Vanuatu, Fiji, Samoa and Tonga – had each acted as a filter. A greater variety of Y-chromosomes arrived at each staging post than left it, as some men stayed behind. By the time the first Polynesian canoes reached Rarotonga, the successive genetic filters had whittled down the variety so much that there were only two clusters of Y-chromosomes left and the men within them, we could tell from the fingerprints, were very closely related. The island filters had worked on women too, which is why Rarotongan mDNAs are all so alike.

However, this clearly hadn't happened with the cluster C chromosomes. They were all totally unrelated to one another. They had certainly not been filtered by the islands. That meant they had arrived either separately or as part of a large colonization involving unrelated men. But from where? A big settlement from South America was a distinct possibility. Unlike the way into the Pacific from the west along island chains, the route into Polynesia from South America is across a vast and empty ocean. Many separate journeys from South America by unrelated men could easily have brought a whole variety of Y-chromosomes to Polynesia, as we had found in the members of cluster C. However, if that were the explanation, the South Americans could not have brought any women with them. There were no signs of Amerindian mDNA anywhere in Polynesia.

At the time we did this research we could not tell whether our Polynesian cluster C Y-chromosomes came from South America or not. These were still early days when the paternal family tree of our species was only beginning to be drawn out. It would soon be refined into its modern form, illustrated earlier in the chapter as figure 3, but in 1998 when we did this piece of research, Y-chromosomes from many very widely separated parts of the world were clumped together into the same cluster. Cluster C was one such. Cluster C chromosomes had been found in America, all over Asia and also in Europe. We could neither prove nor disprove the American origins of our Rarotongan cluster C chromosomes. Matt tried to narrow down the possibilities by running the samples through a second fingerprinting system recently developed in Leicester. The characteristics of the Rarotongan chromosomes after they had been through Matt's new system eliminated Asia as the origin. So they were either from America or from Europe. Between those two possibilities, even the new method could not make a distinction.

The solution to this frustrating dilemma emerged when a US research team found a new genetic marker which was finally able to distinguish American from European cluster C chromosomes. As soon as we heard about this, we rushed to test our ten mystery chromosomes – with decisive results. The cluster C chromosomes which we had found in Rarotonga, which made up nearly a third of the total, were definitely not from South America; they were from Europe! Almost a third of the Rarotongan men we had analysed had inherited their Y-chromosome not from one of the original settlers but from a European man. This was such an extraordinary result that we could scarcely believe it. But there was no doubt. These Y-chromosomes had come from Europe. We had never seen a single mDNA from Europe anywhere in Polynesia. From the mDNA evidence alone, it was as if these islands had never been visited by Europeans. But the Y-chromosome told a completely different story. The traces of European men were everywhere.

Knowing the history of Polynesia, this is not difficult to explain. The Europeans who first visited the Polynesian islands were all men. There simply were no female explorers or whalers or sailors or traders or missionaries; so the arrival of Europeans made no impact whatsoever on the mDNA gene pool of modern Polynesians. As for the prevalence of European Y-chromosomes, if we had no idea of the history of the islands we might imagine this to be the echo of a past military invasion led by men, which would produce a large influx of Y-chromosomes but not of mDNA. But that isn't quite what happened in Polynesia. European men often needed little persuasion to jump ship, and in most cases they were made welcome. In the Cook Islands at least, it was not uncommon for mothers to encourage their daughters to marry European men, even bringing them from the outlying islands to Rarotonga specifically for that purpose.

Polynesia changed completely after the arrival of the Europeans. New infectious diseases swept through the susceptible islanders. Social structures crumbled as the old religions were swept away by aggressively evangelical missionaries. It is a familiar story, repeated time and again across the world. But suspend your disapproval for a moment to look at what we had found from a purely genetic point of view. Consider the European arrivals as a case of boy meets girl – or, more specifically, of sperm meets egg. The European ships brought sperm from the other side of the world to fertilize the eggs of Polynesian women, rather than the other way round. There were no European eggs on board – they had all stayed at home. Was sexual selection at work here? The European sperm must have found its way to the Polynesian eggs somehow, either by force or by consent. The records of Polynesian mothers bringing their daughters from the outer islands to marry Europeans suggests at least some degree of female choice, if only by the mothers. What had these new men got to offer that their Polynesian rivals hadn't? These men were not rich by European standards at home, but on Rarotonga they had wealth and status on their side and, as a result, their sperm found eggs to fertilize. Female choice, the engine of sexual selection, was selecting features familiar to us all.

What of the genes themselves? Which were the winners and which were the losers in this exchange? That is an easy question to answer. The clear winners were the European Y-chromosomes. They had displaced the original Polynesian Y-chromosomes in a third of Rarotongan men. European mDNA had got nothing out of this trade at all; it was nowhere to be seen. Polynesian mDNA, on the other hand, had neither flourished nor suffered from the arrival of the Europeans. It had just carried on, blissfully unconcerned by the battle being fought between the Y-chromosomes. A genetic cynic might even say that the whole European exploration of the Pacific had been organized for the benefit of European Y-chromosomes.

The genetic effect of European colonization in Polynesia has been repeated in many different parts of the world. Now that scientists are beginning to grasp the advantages of analysing both mDNA and Y-chromosomes together rather than sticking resolutely to one or the other, similar or even greater patterns of Y-chromosome success have been found in several different parts of the world with a history of European colonization. A recent study in Peru among inhabitants of Pasco and Lima who, so they thought, had unmixed Amerindian backgrounds found that while 95 per cent of mDNAs were clearly Amerindian, over half the Y-chromosomes were European. Another study, in the Colombian province of Antioquia near Medellin, found that 94 per cent of Y-chromosomes were European, 5 per cent were African and only 1 per cent were native Amerindian. Antioquia was one of the first Spanish settlements in South America, founded in the early sixteenth century, and the 5 per cent of African Y-chromosomes no doubt arrived via the Atlantic slave trade. When the mDNA from the same men was analysed, 90 per cent was native Amerindian, the rest being a mixture of European and African. The picture was clear. European, and African, sperm had been fertilizing Amerindian eggs on a massive scale, both in Peru and in Colombia. European Y-chromosomes had been the major beneficiaries of these colonial adventures, at the expense of native Amerindian Y-chromosomes which in Colombia had been almost wiped out. But still the pattern of mDNA was relatively undisturbed. The women, for whatever reason, had chosen to mate, or been coerced into mating, with European men. This was clear genetic evidence of some kind of sexual selection operating on an enormous scale.

The record of European colonizations in Polynesia and South America was so clear in the genetic record that I predict the same pattern will emerge wherever large-scale European colonization or exploitation has occurred – in North America, Australia and New

Zealand, for example. In the US I would not be at all surprised to see much higher frequencies of European Y-chromosomes than of mDNA among African Americans whose ancestors were once enslaved. A hint of this has very recently been unearthed in a twin mitochondrial and Y-chromosome study of 200 British men whose parents or grandparents emigrated to England from the Caribbean. The clear signs of an African maternal ancestry were found in the mDNA of 98 per cent of these men, but when the Y-chromosomes were tested a quarter of them were European – the unmistakable genetic legacy of rape and seduction. The genetic winners are the incoming Y-chromosomes; the clear losers are the Y-chromosomes of the original inhabitants or, in the case of the Afro-Caribbeans and African Americans, exploited ethnic groups. If this was the pattern following historical colonizations far from Europe, could I find any record of similar events closer to home? Was the driving ambition of the Y-chromosome to dominate and subjugate all genetic opposition chronicled within the cells of Europeans themselves?

15

BLOOD OF THE VIKINGS

At nine-thirty every evening the Inverness sleeper leaves the charmless wastes of Euston station in London and heads north towards the Scottish Highlands. During the high summer it is still light when the train leaves, and for me one of life's pleasures is sitting in the lounge car, glass of wine in hand, watching the sun slip down towards the horizon and feeling the tentacles of urban life dropping away as the train forces itself ever northwards, breaking their grip. I have got to know this train journey well after deciding a few years ago to continue my genetic research in the north of Scotland. Now that I was able to grasp hold of the ends of the twin genetic threads of maternal and paternal history and had seen how well the combination had worked in Polynesia, I thought I was ready to tackle the genetic legacy of another historical colonization. This time I wanted to focus on Britain at the receiving end of foreign adventure – by a people whose very name is enough to make the blood curdle. The Vikings.

As I travelled around the stunning landscape of the Scottish Highlands, I fell in love with this beautiful country where the mountains meet the sea. Were it not for the temperature and the treeless landscape, the Western Isles, with their long, brilliant white beaches and turquoise sea, would not be out of place in the south Pacific. But this sea is not the usually peaceful ocean that washes the distant shores of Polynesia; it is the fierce and stormy north Atlantic. Sitting at the top of a dune of silver shell sand on the Hebridean island of South Uist, watching the great orange lollipop of sun sinking into the calm summer sea, reflected in the ribbon of light that sparkles its way to the horizon, this is hard to imagine. But how different this scene is in winter. Around me on the dunes lie the shrivelled stems of kelp, ripped from the seabed and flung fully fifty yards inland by the terrible fury of winter storms. The winds shriek in from the west as low-pressure systems file one after the other across the north Atlantic. Huge seas crash into rocky headlands and sweep roaring up the beaches under angry grey skies. Spray blows off the breaking waves and mixes with the sand into a blinding hail that stings every inch of exposed skin. This is the other face of the Atlantic, its usual expression: fierce, unpredictable and very dangerous.

Over a thousand years ago the Vikings crossed this treacherous sea to these islands from the deep fjords of Norway. From the sparse remains of Viking longhouses we know they were here in the north of Scotland, but to what extent they actually settled, and what was the character of the settlement, remains unclear. Though the historical accounts are unanimous in their references to the violence of Viking raids, they are silent on whether the settlements were built up by Viking men who married local women or whether the invaders brought their families with them. As well as settling the north of Scotland, the Vikings pushed further into the wild seas to explore Iceland and Greenland, and they eventually reached North America. By all accounts, the Vikings did not establish a permanent presence in North America and the Greenland encampments were eventually abandoned, but the settlement of Iceland was a triumph. Today this spectacular island, a land of glaciers, geysers and volcanoes, is home to a quarter of a million people whose language and culture have strong and unambiguous Norse connections and whose history is set down in epic sagas. There was no mystery about the origins of the Icelanders that needed to be unravelled through genetics, but I began to wonder what a combined mito-chondrial and Y-chromosome survey might yet reveal. Were the genetic connections to Norway as strong as everyone assumed? More intriguingly, was it Viking families or only Viking men who settled Iceland – and, if the latter, where did their women come from?

If the Viking Age can be said to have had a defined opening, then it was on a summer's day in the late eighth century AD. On 8 June 793 a small Viking fleet landed on the island of Lindisfarne, off the north-east coast of England, and attacked the undefended monastery of St Cuthbert. In the space of a few hours they slaughtered what monks they could find and made off with the rich treasure of the church. Isolated and undefended, the monastery at Lindisfarne had no chance at all. The success of the raid en-couraged a spate of copycat attacks on other vulnerable monasteries around the coast of Britain: Jarrow, down the coast from Lindisfarne, in 794; St Columba's church on the remote island of Iona off the west coast of Scotland in 795, again in 802 and yet again in 806, after which it was evacuated back to Kells in Ireland. These coastal monasteries were attractive targets. They were near enough to the sea for the raiders to be able to launch a surprise attack, they were completely undefended and they were full of holy treasures – gold and silver caskets containing the relics of saints, often encrusted with precious stones; the covers of illuminated gospels, equally lavishly endowed with gems and precious metal. Uninhibited by any deference to the Christian church and answer-able only to their own pagan gods, the early Viking raids were ferocious, bloody and very effective. Over the next seventy years, Viking raiding parties became larger and much more ambitious. At first only Britain and Ireland suffered, but before very long settle-ments all along the North Sea and Atlantic coasts became the targets of repeated and sustained attack.

The Vikings established bases on offshore islands at the mouths of large navigable rivers and used these to launch devastating raids on cities deep inland. In France, Rouen and Paris were attacked along the Seine; the Loire led the Vikings inland to Angers, Tours and Orléans; and the Garonne exposed Bordeaux and towns almost as far east as Toulouse. A vast fleet of 150 ships attacked Lisbon on the Tagus then Seville along the Guadalquivir before entering the Mediterranean, where they made a forward base on an island at the mouth of the Rhône. From here the Viking ships attacked Avignon and other cities on the great river and raided the coast of Italy. These were campaigns in the heroic tradition of their pagan gods, not cowardly raids on undefended monasteries. The Vikings were often beaten back – they lost two hundred men during their attack on Seville, for example – but these reverses served only to increase the heroic tenor of their exploits, from which they returned with those two most valued of male commodities – honour and profit.

But what trigger unleashed the rampaging Norsemen on the rest of Europe? Until the attack on Lindisfarne, the ancient Scandinavian world had kept itself to itself. There was good fishing in the fjords and enough fertile land around the coast for mixed farming of cereals and animals, mainly cattle and sheep, and the forests provided wood for fuel and timber for building. This was a self-sufficient culture and, with a religious tradition dominated by heroic gods and goddesses, one that lay outside the largely Christian world of the rest of Europe. Their environment was certainly harsh but it was also magnificent, and the Scandinavians' system functioned smoothly enough with little outside interference. Why did they suddenly leave this self-sufficient world of fjord and forest, which had been their ancestral home for thousands of years? What was the stimulus that sent the Vikings to every part of Europe in one of the bloodiest periods in the continent's history?

There are conflicting theories surrounding the Vikings' sudden desire to leave their homeland. One suggestion is that the population began to increase, perhaps helped by a slight improvement in the climate. With only very limited amounts of land available for cultivation, there was increasing pressure on space. Farmsteads were passed from father to eldest son, leaving other sons with nowhere to go when the land ran out. Nor did younger brothers lose out on land alone. The men with the land got most of the women, and polygamy was widespread. Thus, for many young men there was every incentive to look elsewhere for somewhere to live – or at least for some other means of getting a woman. It was this most basic of instincts that drove the young men to risk the wrath of the north Atlantic. Adam's Curse had begun to growl, the cruel mistress of sexual selection driving them across the seas to seek the rewards offered to heroes. Men slain in battle stood a good chance of getting into Valhalla – the paradise of the afterlife – if their bravery on earth sufficiently impressed the Valkyries who guarded the entrance. And heroes always get more women. However, although it was the richness and the vulnerability of the monasteries which first drew the Vikings to Britain, it was the prospect of fertile land that could be farmed which was to prove the lasting attraction. The quest for honour and profit was gradually replaced by the search for somewhere to live.

As the aggressive Viking war fleets were plundering the coasts and inland rivers of Britain, France and Spain, other Norsemen were beginning to settle on the islands of the north Atlantic: first Shetland, the land nearest to the coast of Norway; then Orkney to the south, then Lewis and the Uists in the Western Isles. Here was land to farm, the commodity in such short supply at home, set in the same familiarly wild and spectacular environment. The peculiar fertility of Orkney soon established this emerald-green archipelago as the political centre of Norse influence and within a hundred years the earldom of Orkney controlled not only Orkney and Shetland but the north and west coasts of Scotland, the Western Isles and the Isle of Man.

But how and by whom were these settlements established? There is abundant archaeological evidence that the Scottish islands had been settled for at least five thousand years before the Vikings arrived. For the most part the islanders lived in isolated family settlements, often centred on fortified round houses made of stone. Did they suffer the same fate as the monks at Lindisfarne, put to the sword and their farms taken over? There is actually very little archaeological evidence for a violent ousting of the indigenous farmers. There are none of the tell-tale signs of charcoal or scorched stone that are the usual witnesses to massacre and fire. The characteristic long rectangular houses of the Vikings are sometimes built away from the Pictish round houses, while at other archaeological sites there is a gradual transition, with buildings of both styles merging one into the other. But while the physical evidence for slaughter is absent, it does not mean that this was an entirely peaceful settlement. Bearing in mind the violence of Norse adventures in other parts of Britain and Europe, it would be surprising if the Viking settlers on the Scottish islands spent their evenings engaged in small talk with their new neighbours who had invited them in for the eighth-century equivalent of a dry sherry.

I was becoming more and more convinced that the parallel examination of mitochondrial and Y-chromosome DNA would be able to tell us a great deal about both the nature and the extent of the Viking settlements. The challenge had many of the same ingredients as my research in Polynesia, but with additional difficulties. There it had been easy to tell the difference between Polynesian and European genes; for the same approach to work in Scotland, we would need to be able to distinguish Norse mDNA and Y-chromosomes from those of the original Pictish inhabitants. Since both were essentially European and presumably shared a common ancestry not so very long ago, it might be hard to tell them apart. Mutations in DNA accumulate with time, so mDNA and Y-chromosomes with a recent ancestry in common will be more alike than those which are only distantly related. I am, of course, talking about vast lengths of time. An ancestor might have been dead for ten thousand years and still qualify as recent!

By a stroke of luck, as I was halfway through this piece of research an Icelandic anthropologist, Agnar Helgason, arrived in Oxford to complete his PhD on the genetics of his own island. He had already collected DNA from his compatriots and we soon agreed to compare results. By then my research team and I had made dozens of trips on the sleeper and collected DNA from thousands of volunteers from all over Scotland. The difficulty which Agnar and I faced was how to distinguish Icelandic genes that had come from Norway from those that had arrived from elsewhere. As Iceland was uninhabited when the Vikings arrived, we didn't have to worry about any original inhabitants there, but in Scotland we had to be able to tell Norse genes from Pictish. Fortunately we already knew quite a lot about Norway. Jayne Nicholson, who had done the Y-chromosome work in Polynesia, and another researcher, Eileen Hickey, had anticipated the need for Norwegian DNA samples and organized a visit to the main blood-transfusion centre in Oslo. They were rewarded with hundreds of samples from donors coming from all over Norway.

When we all met up to begin the comparisons, we wondered whether we could take the DNA result from each *individual* and estimate whether his or her ancestors had been Vikings or not. We looked out the first Icelandic mDNA sequence from Agnar's own results. I copied it onto my laptop and searched our Scottish and Norwegian results for a match. Within a fraction of a second the computer had found exactly the same sequence in three Norwegians but not in Scotland. This was a promising start, though it didn't mean that this particular mDNA is present *only* in Iceland and Norway. To know that for certain we would have to have the DNA sequence from everybody on the planet. But it did make it extremely likely that the maternal ancestor of our first Icelander was indeed a Viking woman. The mDNA sequence from the second of Agnar's Icelanders, put through the same process, exactly matched one sample from Ireland which we had analysed years before, and two more from the Grampian region in north-east Scotland, but none at all from Norway. We scored that as non-Viking. Our third Icelander had an mDNA that matched two people from inland Scotland, two from Ireland – and four Norwegians. It was bound to happen. Was this a Viking or not? We just couldn't tell.

After several cups of coffee we decided to score this, and other ambiguous sequences that we came across, as intermediates. As we went down the list one by one, we refined the system so that each individual was scored for their Viking or Irish/Scottish (which we eventually called 'Gaelic') mitochondrial affinity. People whose mDNA match was found in Norway but not in Scotland or Ireland were given a score of 100 per cent Viking and 0 per cent Gaelic. Individuals, like Icelander number 2, with matches in Scotland and Ireland had the mirror-image score of 0 per cent Viking, 100 per cent Gaelic. Icelander 3 scored as 50 per cent Viking, 50 per cent Gaelic, because the matching sequences were equally common in both Norway and Scotland. Sometimes we found Icelanders whose mitochondrial DNA matched up with a sequence that was present, but rare, in Scotland but common among the Norwegians. Depending on the exact figures, these Icelanders might be given a score of, say, 90 per cent Viking and 10 per cent Gaelic. That is not to say that 90 per cent of their mDNA came from Norway and the other 10 per cent came from Scotland. Their mDNA can't possibly come from *both* locations. It just means that we estimated the chance of its being Viking as 90 per cent and the chance of its being Gaelic as 10 per cent. Occasionally we would come across Icelanders whose sequence didn't match either Norwegian or Gaelic samples; in these cases we found the closest match we could, figuring that the DNA had mutated since it arrived in Iceland or that we simply hadn't encountered the exactly matching sequence in either Scotland or Norway. When we had been down the complete list, we added up the figures in the percentage Viking and percentage Gaelic columns and divided by the number of individuals to get the overall figure. It came as a complete surprise. By this count, the Icelanders were more Gaelic than Viking! By our estimation 60 per cent of the Icelandic mDNA had a likely Gaelic ancestry and only 40 per cent seemed to have come from Norway.

It was obviously important to do the same calculation for the Y-chromosome. If this gave similar results then we had to begin asking ourselves searching questions about our methods. Just as we had done for mDNA, we went through the Icelanders one by one, matching them to Y-chromosomes from our Gaelic and Norwegian volunteers and giving each Y-chromosome a score which reflected its likely ancestry. When we summed the figures I felt a palpable sense of relief. By our test 70 per cent of the Icelandic Y-chromosomes had a Viking ancestry with the remaining 30 per cent having a Gaelic origin. The proportions were reversed. The Y-chromosomes of the majority of Icelandic men had been passed down in direct paternal descent from the Vikings. On the other hand, most Icelanders had inherited their mDNA not from the Vikings but down maternal lines tracing back to Scotland and Ireland.

The Norse settlement of Iceland began in about 870, nearly eighty years after the attack on Lindisfarne that heralded the dawn of the Viking Age. The settlement reached its peak at the beginning of the tenth century, at a time when Viking fortunes elsewhere in Europe were on the slide. They had been beaten back from their estuarine strongholds on mainland Europe, while the Great Army which had terrorized England had been weakened by the West Saxons under King Alfred and gradually dissipated. In Ireland, where they had never really managed a sustained settlement, the Viking strongholds were gradually destroyed by the numerous Irish kings, a process culminating in their expulsion from Dublin. Almost everywhere, it seemed, the Vikings were on the run. Only in Normandy had they managed to settle permanently, after striking peace deals with the inland rulers. The discovery of Iceland, empty, fertile and surrounded by fish, couldn't have come at a better time. There were no prospects for the roving Vikings back in Norway, where all the available land had been claimed long ago. Besides, a succession of land-grabbing kings were making life uncomfortable for even those Norsemen who had stayed at home. A lot of them wanted to leave as well. There was a rush for the new land. But by whom?

The genetics ruled out a straightforward migration of Norse families direct from Norway as the only source. Had that been the case then we would have found most of the Icelandic Y-chromosomes *and* mDNA having their closest matches in Norway itself. Had the first Icelanders come straight from Norway, bringing their wives, we would have seen roughly equal numbers of Norse mDNA and Y-chromosomes. As it was, less than half the Icelandic mitochondria started their sea voyage from Scandinavia. Less than half, but substantially greater than zero. So we could also rule out a mass migration of Viking men, picking up wives on their way, as the only settlers in Iceland. Did anybody at all go to Iceland direct from Norway, or was the great settlement made up entirely of Vikings on the run? To find the answer to this question, we set about doing the same analysis that we had done for Iceland on the parts of Scotland that came under the influence of the Vikings – Shetland, Orkney and the Western Isles. Once again we went through the individual results one by one and assigned each of them to either a Viking or Gaelic source, using our results from parts of Scotland which had never been under Norse control as our yardstick for 'Gaelic' ancestry. Plotting the Y-chromosomes first, we found that 35 per cent of Shetlanders had a Viking ancestry. There were slightly fewer (32 per cent) in Orkney and fewer still (24 per cent) in the Western Isles.

These results told us at once that neither Shetland nor Orkney had been completely over-run by Viking males. Norse cultural domination might have been complete in these islands – for example, no Pictish place names remain – but only just over a third of the men living there now have a Viking paternal origin. Of course, we are reconstructing past events from what we see in the modern population, and a lot has happened in the intervening centuries, including substantial immigration into Shetland from the Scottish mainland after the islands were finally returned to Scottish rule in 1472. Even so, these results are not telling us that the initial Viking settlements killed or displaced all the original Pictish inhabitants. A substantial proportion were left alive to pass on their Y-chromosomes to the modern-day men of the islands.

What of the women? When we repeated the procedure with mDNA, the results were completely unexpected. While we were slightly surprised that the proportion of Norse Y-chromosomes in Shetland and Orkney had not been higher, the mDNA figures were astonishing. There were as many mitochondria of Scandinavian descent on Shetland and Orkney as there were Y-chromosomes. That meant, taking into account the general proviso about reconstruction from modern data, that as many Viking women had settled there as had Viking men. The Vikings had come as families. This had to mean that, at the same time as the ferocious Viking war fleets were marauding and plundering the coastline of mainland Britain, other Norsemen were settling their families on Orkney and Shetland in comparative peace and quiet. We had expected to find the proportion of Viking mitochondria in the islands to be far lower than that of their Y-chromosomes, imagining from their reputation elsewhere that the Vikings would have killed the men and taken their wives. But we were wrong. They must have brought their own women with them.

Further west, in the Western Isles, there are fewer Viking Y-chromosomes. About a quarter of men have a Viking paternal ancestry, and this is no surprise given the reduced influence the Vikings exercised here compared to Orkney and Shetland, which were much closer to Norway. Even so, a quarter is a substantial proportion. But the number of Viking mitochondria, signalling Viking women who settled, is very much smaller. Only 8 per cent of the mDNA in the Western Isles today has a Viking origin by our tests; the remaining 92 per cent is of Gaelic ancestry. So some women did come here from Norway, but not many. In these islands we have the familiar signal of male-dominated settlement that we have already seen in Polynesia and South America, where the new settlers took local wives rather than bringing their women with them.

What does all this say about Iceland? The genetics eliminated an entirely family-based settlement direct from Norway – there are far too many Gaelic Y-chromosomes and mitochondria in Iceland for that. I think the likeliest explanation from these results is that most of the original settlers came to Iceland from Norse settlements around the coast and offshore islands of Scotland and Ireland, places which had already experienced one or two generations of intermarriage of Viking men with Gaelic women. The presence of so many Gaelic Y-chromosomes in Iceland also suggests that the intermarriages had gone both ways in the Norse settlements, with Gaelic men marrying the daughters of Vikings. That is the kinder explanation of the genetic results. The other is that Gaelic women, and men, were taken to Iceland as slaves – the men to work the fields, the women to breed. I do not want to exaggerate the accuracy of what genetics tells us about the Vikings. It cannot on its own reconstruct a complete picture of what went on in the past – it can only contribute to it. The other elements of archaeology, linguistics and written history, each hedged about by their own provisos, premises and uncertainties, are just as important. But what the combined genetic approach, which separately traces the history of men and of women, has done both in Polynesia and in the islands of the north Atlantic is to illuminate aspects of our human past that were hitherto hidden from view.

The Age of the Vikings has all the hallmarks of Adam's Curse: the insistent urge of men to mate with as many women as possible, and the intense rivalry among Y-chromosomes that ensues. As their first-born sons accumulated wealth enough to collect women at home, their unfortunate younger brothers, dispossessed of the means to attract a mate as surely as if they were peacocks with their tails trimmed, set off across the seas to look for sex on distant shores. When they found land, some returned to Norway to claim their prize and took her back to the new colonies. Their tails had regrown. Others did not bother to go back to Norway and settled with native women. The record of their success is carried down to this day in the Y-chromosomes and the mitochondria of the men and women who still live in these wild and beautiful islands on the edge of a furious sea.

16

THE Y-CHROMOSOME OF SOMHAIRLE MOR

Each piece of mitochondrial DNA, each Y-chromosome, has its own story to tell, of a battle fought long ago, of a heroic journey in times past as our genes flow through ancient time to their present guardians – you and me. Now we can follow their journeys to the jagged coasts of the north Atlantic or to the soft coral sands of the south Pacific. These tiny pieces of DNA, each the separate ambassadors of the essence of the feminine and the masculine, have travelled to these distant lands just as surely as the longship and the outrigger carried their temporary custodians – the bodies of our ancestors.

These are the stories of the travels of our genes just as much as of the adventures of our ancestors. What naked force drove them into the unknown, across seas churned by maelstrom and cyclone to the lands beyond? For the Vikings I have already offered a conventional motivation: the overcrowding and shortage of land at home, the grim prospects for younger sons, and the greed of ambitious kings coupled with the means of escape. Their Y-chromosomes had no future in Norway, so they had to get away – and they did. They also carried their motivation with them. How proud the Shetlander who discovers that he carries the chromosome of the brave Viking who first hauled his boat onto the golden beaches of Yell or Uist! Within each cell he carries the evidence of his pagan and heroic past. He also carries the motivation – his Y-chromosome. That tiny shred of DNA and its ambition to survive and multiply is what launched the ship of his ancestor from the deep fjords of Norway into the setting sun. His Y-chromosome drove him into the giant waves of the north Atlantic, sensing that its future lay beyond the horizon. It will not have particularly cared whether it propagated itself with the assistance of Norse women or the females that he knew could be fought for at the journey's end. Sometimes it paid to take your own women along, sometimes it didn't. To the Y-chromosome it is a matter of complete indifference. The important thing was to get away, to escape extinction by the ambition of other Y-chromosomes, particularly the king's, and to survive. If this meant killing another man for his wife, then the Y-chromosome would be indifferent to the pain and despair. Survive and multiply. That's all that mattered.

As I thought about this I began to wonder whether some Y-chromosomes are better at multiplying themselves than others. Were there Y-chromosomes that had multiplied far more than their contemporaries? If the world is being shaped so much by the ruthless ambition of the Y-chromosome, as research in Polynesia, in South America, in the Caribbean and among the Vikings was certainly suggesting, was the spell cast more effectively by some men than others? The answer turns out to be very surprising. And, like many surprises, it had very prosaic beginnings.

During my search for Viking Y-chromosomes in the Highlands and Islands of Scotland, my research team and I had collected together several thousand DNA samples. One of the first things we always do with DNA results is to draw out an evolutionary network so that we can see how the different DNA sequences, whether from Y-chromosomes or from mDNA, are related one to another. It is just this kind of treatment which revealed that the mDNA of most native Europeans fell naturally into seven quite distinct clusters. The same kinds of evolutionary networks can also be constructed for Y-chromosomes and, once again, a number of different clusters appear. In Britain, it happens that the great majority of Y-chromosomes fall into one of three clusters. One of them is the third cluster that we had seen in Rarotonga, the 'cluster C' chromosomes which had their origin on the other side of the world in western Europe. These are generally referred to as 'class 1' chromosomes. The other two clusters which are found widely in Britain are the class 2 and class 3 chromosomes. It is no coincidence that these clusters occupy the first three numerical slots in the list, for the system of markers which distinguishes them was devised by Mark Jobling and his colleagues – who live and work in Britain.

Y-chromosomes in different clusters are not closely related to each other, so before we even began to draw networks from our Scottish results in detail we separated the Y-chromosome results into their respective clusters using the system devised by Mark Jobling and his team from Leicester. This done, we next mapped out the detailed evolutionary relationships within each cluster, using the highly variable genetic fingerprinting system based on DNA repeats that Jayne Nicholson had developed for our Polynesian research. This exquisite system is capable of distinguishing about half a million different Y-chromosomes. These Y-chromosome fingerprints are superb indicators of recent genetic ancestry – by which I mean within the last thousand years or so. Within that time range, there is a good chance that identical Y-chromosome fingerprints are inherited from a common paternal ancestor. It is just this logic we used to track the ancestry of Icelanders to their Norse or Gaelic origins – and also to prove the common ancestry of so many Mr Sykeses.

As Jayne and I plotted out the three classes of Scottish Y-chromosomes one day, we were both struck by the uneven distribution of the various fingerprints within each of the three clusters. I was very accustomed to looking at mitochondrial networks where, by and large, the different DNA sequences within each cluster are related to one another in a sensible way that I will endeavour to explain using figure 4. Part (a) of the diagram shows a typical example from one of the European mDNA clusters. Each of the circles represents a particular mitochondrial sequence that we have found and we drew them so as to be proportional in size to the number of people who have each sequence: the bigger the circle, the more people have it. The circles are joined together by lines which themselves represent the difference between the two sequences. Here, the longer the line, the bigger the difference between the sequences, and vice versa. These differences in DNA sequence are caused by mutations, so two circles with only a single mutation separating them are joined by a short line while those separated by more mutations are connected by proportionately longer lines.

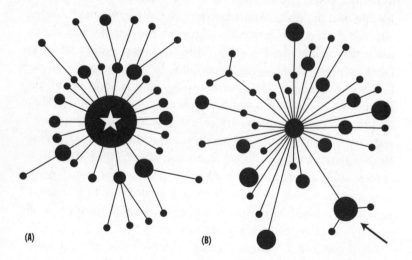

(A)   (B)

Figure 4: Contrasting cluster patterns in (A) mitochondrial DNA and (B) Y-chromosomes

The big circle with a star is the ancestral sequence, still inherited unchanged by a large number of people from their common maternal ancestor, the clan mother. The small circles radiating from this central core are mDNA sequences, shared by far fewer people, which have experienced one or two single mutations away from the clan mother's sequence somewhere in their maternal genealogy. Occasionally an even smaller circle branches away from one of these, representing yet another mutation. But there is a definite pattern. The largest circle is always the one in the middle. It is the sequence of the clan mother. This is always larger than the circles at one mutation away, which are themselves mostly bigger than the circles at two mutations distant from the ancestral sequence. This makes complete sense as the mutated sequences, being more recent, will have progressively fewer descendants than the older ancestral sequences. Eventually the pattern breaks down over tens of thousands of years as all the maternal genealogies coming down from the ancestral mother experience at least one mutation and the central circle gradually disappears.

I was expecting to see the same sort of pattern with the Y-chromosomes in each of the clusters, with a large common central signature orbited by circles representing Y-chromosomes at one, two or more mutations distant. However, that was definitely not what we saw, as the example in figure 3(b) shows. The networks were straggly and ragged. Although we did eventually put one circle at the centre, it was by no means obvious which we should choose. Frankly, there were no obvious central circles, no plainly ancestral signatures. Here and there, in a pattern that lacked any consistency, a large circle stood out with one or two orbiting satellites. In other parts of the network, single signatures, claimed by just one individual, would appear on the network quite un-related to anything else. What was going on?

I was well aware that the mutation characteristics of the Y-chromosome fingerprinting system were erratic, with double or even triple mutations known to happen at one go. Perhaps that was the explanation for the bizarre appearance of the network, though it was hard to see why. These double or triple jumps would have been very rare events, not expected to disturb the majority of Y-chromosome signatures as they diffused away from the central ancestral sequence one step at a time. I did not understand the pattern in the networks and I am ashamed to say that I put it out of my mind – ashamed because it is precisely those things that you don't expect, those results that do not fit with your expectations, which can lead towards a new discovery. Results that turn out as predicted are of course important, but they rarely do more than add one more layer to the existing piles of knowledge rather than start a completely fresh one. But there it is: I did ignore this ragged network and my mind relapsed to concentrating on the hundred and one other things that cried out for attention.

At our next meeting, Jayne produced her latest version of the unruly networks. But this time she had brought along something else as well: a list of surnames. Because we had collected all the samples ourselves, we knew precisely who they belonged to. This was around about the time that I had begun to think more about surnames after my surprise results with the Sykeses. Jayne had gone back to the files where we stored the names of our volunteers and had extracted all their surnames. Most of our Scottish volunteers had given us their DNA at blood-donor sessions, so there was no reason to expect any names to be especially common in our collection, unless they were generally frequent in the areas we had visited. With the list of surnames in one hand and the ragged networks lying on the table in front of us, we started to compare them.

Jayne had already taken a preliminary look and pointed out one particular circle which stood out. I have marked it with an arrow on the right-hand side of figure 3(b). Looking down the list of names, we could see we had found the particular Y-chromosome fingerprint represented by this circle in five men with the same surname – Macdonald. Two were from North Uist in the Western

Isles, one from Skye, another from the Borders and one from near Inverness. Macdonald is the commonest name in the Scottish Highlands, so it was no surprise that we had encountered quite a few on our travels. We also found this unusual chromosome in a Mr Barclay from Shetland, a Mr Ferguson from Argyll, a Mr MacAlister from the Isle of Mull and a Mr MacDougall from Glasgow. There didn't seem all that much in it until Jayne said quietly, 'You know the Macdonalds, MacDougalls and the MacAlisters are all supposed to be related, don't you?' I didn't. Not then. But I do now.

The possibility, remote as it was, that this Y-chromosome was inherited from the common ancestor of the Macdonalds, the MacDougalls and the MacAlisters was incredibly exciting. Although by then I had begun to realize that most Mr Sykeses shared a common ancestor, I never imagined for a moment that we would ever find anything remotely similar among the great Scottish clans. I was familiar enough with Scottish history to know that it had once been the custom for crofters and tenant farmers to adopt the name of their clan chief. That would so fatally confuse the relationship between surname and Y-chromosome that I had not entertained the thought, even for a second, that there might be an identifiable genetic connection among members of the same clan. While we were pressing on with a study of several other English surnames, with similar or even more impressive results than we had obtained from the Sykeses, it had never seemed remotely conceivable that we could do the same in Scotland. But there was Jayne's evidence. It was certainly slim, but definitely worth pursuing.

Jayne wrote to dozens of Macdonalds, MacDougalls and MacAlisters from all over Scotland. In her letter was an invitation to use the small sampling brush she had enclosed to remove a few cells from their inner cheeks and return the sample to us. Within two weeks we had more than fifty replies. Jayne set to work extracting the DNA from the cells retained on the bristles, then going through the process of obtaining their Y-chromosome signatures. I, on the other hand, made straight for Euston and the Inverness sleeper. I was going to the Isle of Skye, heartland of the Macdonalds.

Next morning, at Inverness, I hurried over the platform to catch the small connecting train that would take me to Kyle of Lochalsh on the threshold of Skye itself. On a sunny day like this one, it is the most beautiful train journey in the whole of Britain, threading through forest and wide open strath to the village of Achnasheen, where the road to the incomparable Loch Maree heads for the north. From Achnasheen, the track descends Glen Carron and follows the wild river at a distance as it plunges through the steep pine-clad gorge that leads to the rhododendron groves of Achnashellach. Today the mountainside all about is awash with lilac in the morning sun as the train eases out from the dark corridors of leaf. Soon the train reaches the sea at Loch Carron and slowly inches round the southern shore between the high cliffs and the sea. In October 2001 this section of track was destroyed by a landslide which came roaring down the mountain during a heavy storm and into the sea, nearly taking the train with it. But today, before the disaster, there is nothing to stop the steady progress of the little train as it pulls round the corner near the village of Plockton to reveal for the first time the misty hills of Skye on the horizon.

A few minutes later the train pulls in to its final destination, the frankly drab town of Kyle – a mess of grey houses, grey shops, grey garages and a few grey ships. Even on a beautiful day like today Kyle looks as though it is expecting rain. I get into the hire car and drive over the new bridge and onto the Isle of Skye, Eilean a' Cheò, the Isle of Mists. Away to the right, across the wide expanse of Broadford Bay, the rotting cliffs of Trotternish at the north of the island stagger into the sea in a jumble of broken pillars and fallen rock. At the northern tip of Trotternish, within easy view of the Western Isles across the Minch, the ruins of Duntulm, the ancestral home of the Macdonalds, perch on a headland high above the black sea. No Macdonalds live there now; ousted by their perennial and bloody rivalry with the MacLeods of Dunvegan, the Macdonalds were banished to the finger of land that makes up the peninsula of Sleat. I am heading for the last stronghold of the Clan Donald at Armadale on its southern tip – as far away from Dunvegan as it is possible to get and remain on Skye.

Most parts of Skye are gaunt and bare but Sleat is clothed in woods, today filled with the white flowers and smooth green leaves of wild garlic. I can smell them as I wind down the window. Deeper among the trees the ultra-violet mist of bluebells, long over in the south, still glows in the half-light. On my left is the Sound of Sleat where once, sitting with my son Richard looking for otters among the kelp, in the silence we heard the faint but unmistakable draw of the bagpipes. There was no-one to be seen. The wild shore of Knoydart across the sea was at least five miles distant but there was no doubt the sound of the pipes was coming from that direction. Even my strong binoculars could not pick anyone out on that distant shore, unreachable by road. The sound came and went with the breeze – but there was no piper. Richard and I had looked at each other as if to check we were both hearing the same thing. It must be a ghost, a fairy piper long drowned in the sea. It was not scary in the least; it seemed perfectly natural for a ghostly piper to be playing in this wild place. We just sat and listened. I scanned the other shore once more and there, in a tiny sailing dinghy, I saw the tiny figure of a man, the chanter to his lips, moving slowly down the sound towards Mallaig. It may not have been a ghost after all – but it was magic all the same. This time, I stop the car at the same spot and clamber down to the rocky shore, ears straining for the faint sound of the pipes. I scan the sea from the lighthouse of Camusfearna in the north to the white sands of Morar in the south, but no ghostly pipers sail the sound today.

Inside the gates of Armadale, the castle itself is no longer occupied but the estate, now owned and run by the Clan Donald

Trust, is a magnet to Macdonalds from all over the world. The study centre in the grounds houses every detail of the clan history. To Armadale arrive Macdonalds from Canada, the USA, Australia, New Zealand to search for the records of their ancestors. I cannot find an accurate figure for the total number of Macdonalds throughout the world but the chief archivist at the centre, Margaret Macdonald, does not dispute the figure of between three and four million. After laying out our results and giving the briefest of tutorials in the genetics of the Y-chromosome, I asked Margaret whether it was conceivable that we had stumbled across the chromosome of the founder of Clan Donald. This would be present only in those Macdonalds directly descended through the male line – and there were at present only five men who claimed to be so descended, the current clan chiefs.

Together Margaret Macdonald and I settled down to look closely at the clan genealogy. Above their coats of arms, the names of modern clan chiefs were emblazoned across the pages: Ranald Alexander, 24th Chief of Clanranald; Aeneas Ranald Donald, 22nd Chief of Glengarry; Sir Ian, 17th baronet and 24th chief of the Macdonalds of Sleat; and, in the centre, Chief of the name and arms of Macdonald, Godfrey James, Macdonald of Macdonald, 8th Lord Macdonald. From each of the modern chiefs, black lines climbed up the page, coalescing one by one in the names of their common ancestors – Hugh of Sleat, Ranald of Clanranald and John, Lord of the Isles. Deep in the genealogy, the lines were joined by Alastair Mor, the first chief of Clan Alastair of Loup, then higher still by Dugall, first Chief of Clan Dugall of Lorne. This is where the MacDougalls and the MacAlisters claim connection to Clan Donald. Above them all, at the very top of the genealogy, all lines converge on one man – Somerled of Argyll. Is it possible, is it conceivable that we had discovered the genetic signature of Somerled himself? Somhairle Mor: the man responsible, according to legend, for driving the Norsemen from Scotland's western seaboard and reclaiming the land for Gaeldom. Had we picked out from the straggling network of interconnecting circles the Y-chromosome of the man who was without dispute the greatest leader of Gaelic Scotland? If we had, it was doing very well. What was the secret?

What is known of Somerled himself? Here we enter the dappled world of myth and legend, of fact and fiction, where written sources differ according to the inclination or allegiance of the chronicler. That Somerled lived and died there is no doubt. He was born about 1100, the son of Gillebride, whose lands in Scotland had been taken over by the Norsemen. According to clan legend, Gillebride returned to Ireland to ask for help in winning back his inheritance. This tradition places Somerled's father as a direct descendant of a long line of Irish kings going back as far as the second-century Conn of the Hundred Battles. This is, of course, a fitting genealogy for a great Celtic hero. When the Irish Celts began to establish themselves in Kintyre and Argyll in the west of Scotland in the sixth century as the Kingdom of Dál Riata, Somerled's ancestors were among them, but how they came to lose their lands to the Norsemen is not recorded.

A host of tales surround the brave exploits of the young Somerled, 'a well-tempered man, in body shapely, of a fair piercing eye, of middle stature and of quick discernment', according to the clan histories. In one story, after the King of Norway had ordered an invasion of Morvern, on the mainland south of Skye, the in-habitants, apparently leaderless and with the invasion fleet in sight, agree to make the first person to appear their commander. On cue, Somerled appears with his bow, quiver and sword, takes command and fools the enemy into thinking they face a much larger force by marching his men three times round a hill. That done, he leads the charge to the beach, where he slays and rips the heart out of the first warrior he encounters. The Norsemen, overwhelmed by the ferocity of the attack, retreat to their boats and the people of Morvern are free at last. I can see Ewan McGregor in the part already.

After these and other early victories, Somerled continued his heroic campaigns against the Norse in Argyll, Kintyre and the islands to the west until, at last, he appears in the formal historical record in 1153 as the *regulus* or ruler of Argyll. When and how he gained his authority there is not known; nonetheless, Somerled had, by whatever means, established himself as a powerful figure in the west of Scotland and the Isles. Although the understandably favourable accounts portray Somerled as an authentic Celtic hero with an impeccable pedigree stretching back to the ancient Irish kings of Dál Riata, his world was not as sharply divided between Gaelic and Norse, oppressor and oppressed, as his chroniclers no doubt wanted to suggest. It was much more integrated and, as the genetics had shown, there was a great deal of intermarriage between Norsemen and Gaelic women. Even the name Somerled mac Gillebride is a fusion of Norse and Gaelic: the given name, Somerled, derives from the Norse 'sumarlidi' or 'summer voyager', while the surname in unambiguously Gaelic – mac Gillebride – the son of Gillebride.

This fusion of Norse and Gaelic, reflected even in the name of Somerled himself, is probably a more helpful image in thinking about that wild region than one of eternal struggle between two immiscible people. That part of Scotland has always had its own character and has always been fiercely independent. It was not formally incorporated under full Scottish sovereignty until 1493, when one of Somerled's descendants, John, fourth and last 'Lord of the Isles', formally surrendered the lordship to the Scottish king James IV. With an affinity to the seafaring tradition of their Scandinavian ancestors running strongly in the Norse–Gaelic fusion of blood and culture, the people of the far west and the Isles were continually at odds with the central authority of the Scottish kings. Somerled was no exception, and was embroiled in plot and counter-plot against the ruling dynasty.

Just as the Clan Donald histories portray Somerled as an un-diluted hero in his struggle against the Norse, so the chronicles of the Scottish court refer to him as a treacherous rebel, continually betraying the allegiance he owed to his natural lord, the King of Scotland. After backing two unsuccessful rebellions against the Scottish crown and so failing to increase his influence on the mainland, he turned his attention to the Isle of Man. Lying midway between Ireland and the Cumbrian coast of north-west England, the Isle of Man was used by the Vikings as a staging post for their attacks on Ireland before and during the settlement of Dublin in the mid-ninth century. It had long been a Norse stronghold ruled over by a king who derived his authority directly from the kings of Norway. Somerled, during his rise to power, had married Ragnhilda, the daughter of Olaf, King of Man and the Isles – once more demonstrating the intimate mingling of Norse and Gael that was the signature of the western people.

Olaf's son Godred was an unmitigated tyrant and it was not long before a deputation of local chieftains approached Somerled to ask for his help in getting rid of their oppressive ruler. The plan was to replace him with Somerled's eldest son, Dugall, whose claim to the throne of Man was through his mother, Ragnhilda, Olaf's daughter and Godred's half-sister. The attempt to oust Godred reached its climax in a sea battle fought off the coast of Islay in January 1156. According to the Chronicle of Man the battle was fought to a stalemate through the long winter night, with equal slaughter on both sides. When dawn broke, with no decisive winner, Somerled and Godred decided to split the kingdom. Kintyre and Argyll on the mainland went to Somerled, as did the islands of Jura, Mull and Islay, while Godred retained the Isle of Man, the Western Isles and Skye. But Somerled reneged on the agreement and two years later, attacked the Isle of Man, drove Godred out and seized the remainder of his possessions.

Never one to call it a day, Somerled launched a full-scale invasion of Scotland in 1164. He assembled a fleet of 160 ships together with fighters from his own lands and from the Norse enclave in Dublin. The plan was to launch the invasion from the

Clyde at Renfrew on the western outskirts of Glasgow. But this was one campaign too far; it ended in defeat and Somerled was killed. Where he was buried remains a mystery, but the evidence favours his interment on Iona, and certainly this holy site did become the burial place of his descendants. Whether or not Somerled's bones still lie beneath the ground on that windswept island on the edge of the western sea we will probably never know.

The more I read about Somerled, the more I wanted to find his Y-chromosome, the genetic definition of his masculinity. It did not matter that the whereabouts of his own body were not known. He had passed on his Y-chromosome to his male descendants. Even now they, whoever they may be, hold in their own cells the very same fragment of DNA that lodged unseen within the body of the great warrior – Somerled, Somhairle mac Gillebride, King of the Hebrides and Kintyre, Regulus of Argyll, *Rex Insularum*, King of the Isles.

In the search for Somerled's genetic legacy we were looking for a Y-chromosome that was shared among the three clans of Donald, Dugall and Alister, whose own histories linked them back to Somerled. By the time I returned from Skye, now tutored in Clan Donald genealogy, we had heard back from nearly a hundred people with the name Macdonald, MacDougall or MacAlister – and they had all enclosed the small brush that held their DNA. Jayne immediately put the samples through the genetic analysis and we sat down to go through the results. We first divided the Y-chromosomes into the three classes found within Britain before looking at their detailed genetic fingerprints. Beginning with the class 1 chromosomes, we laid out the fingerprints as rows on a spreadsheet and moved them up and down to place identical signatures next to one another. We found six chromosomes that matched exactly, four MacDougalls and two Macdonalds – but no MacAlisters. Another six chromosomes also matched. This time it was five Macdonalds and only one MacDougall – and still no MacAlisters. Another block of six was just the same – Macdonalds and MacDougalls but no MacAlisters. Each time we found a block of identical Y-chromosomes we punched the details of the genetic signature into our database to see if we had seen it elsewhere in Scotland. Each time we did this we came up with several matches from men with a range of different surnames. These were common Y-chromosomes, difficult to distinguish from one another at the resolution we were using at the time. There were certainly structures within this cluster of chromosomes, perceptible sub-groups with shared variants. They did group together but there was nothing particularly striking about them. And none was found in men with all three surnames. If Somerled's chromosome was among them, we could not see it.

There were only very few class 2 chromosomes and among these none stood out, so we went straight on to the final class – class 3. I began to arrange the detailed signatures in order, just as I had done for the first batch. There were twenty-five chromosomes in this class, a little over a quarter of our total. As soon as we began to align the rows I could see that one fingerprint was identical in row after row. In all, nineteen Y-chromosomes were exactly the same. The other six Y-chromosomes differed from this central chromosome by just a single mutation. They must be very closely related. But was this chromosome shared by all men with all three names? I looked at the column with the surnames. Yes! MacDougalls, MacAlisters and Macdonalds: all were there, all with exactly the same Y-chromosome fingerprint. Could this be it?

There was now little doubt in my mind that we had indeed dis-covered the Y-chromosome of the great Somerled – the man from whom the three clans of Alastair, Dugall and Donald claim their descent. To find exactly the same Y-chromosome in men from all three clans, a chromosome which was otherwise rare in Scotland, convinced me that we had identified the genetic legacy of Somerled himself. There was one more thing to do to be absolutely certain and that was to see whether the five clan chiefs still alive whose recorded genealogies descend from Somerled also shared the same chromosome.

This was a delicate task. What if they all shared a different Y-chromosome? That would simply mean that I was mistaken. It would mean that the chromosome which looked so promising for all sorts of reasons did not belong to Somerled at all. That would be disappointing in the sense that my prediction was wrong but, if all the chiefs shared the same Y-chromosome, even if it was not the one I had predicted, then we would still have found Somerled's chromosome. My greater anxiety was that we might find that one or more of the five clan chiefs did *not* share the same Y-chromosome as the others. That would have to mean that their genealogies were wrong; that, somewhere on the lines between Somerled and themselves, so confidently traced in the Clan Donald histories, there was a mistake. One of their paternal ancestors had been adopted or, alternatively, had not been the biological father of his heir. The way to prepare for this eventuality was to make absolutely sure that the results of each test were kept completely confidential and revealed only to the individual and not to anyone else. In that spirit, I wrote to each of the current chiefs: to Sir Ian Macdonald of Sleat, to Ranald Macdonald of Clanranald, to William McAlester of Loup, to Ranald MacDonell of Glengarry, who had recently inherited the title from his father, and to Lord Macdonald himself. Each graciously replied and, with his answer, enclosed the all-important DNA brush. You will have realized already that there was only one possible outcome – they did indeed all share the same Y-chromosome. Had they not, then, of course, I could not have written about it. And the Y-chromosome which all the chiefs shared was the one I had predicted. There was now no doubt that we had identified the Y-chromosome of Somerled himself.

From Somerled this precious talisman passed to his sons with Ragnhilda. After Somerled's death his eldest son, Dugall, who had been installed by Somerled as King of the Isles following the sea battle with Godred, also inherited control of Argyll and Lorne – the lands around and including the Isle of Mull where even today the MacDougalls are still concentrated. His second son, Ranald, inherited Islay and the Kintyre peninsula, while the youngest, Angus, got a scattering of lands to the north of Ardnamurchan and the islands of Arran and Bute, though all these were later seized by Ranald's descendants. Somerled's Y-chromosome passed through Ranald to his grandson Donald of Islay, the founder of Clan Donald. From Donald, the Y-chromosome passed first to his two sons: Alastair, the founder of Clan Alastair of Loup, and Angus Mor. Through Angus Mor, the same chromosome was handed on to all branches of Clan Donald – and all five living chiefs still carry Somerled's Y-chromosome in their cells to this day. The chromosome which was there when Somerled slew the Norse on the beaches of Morvern. The chromosome which was there when he fought the sea battle with Godred of Man. The chromosome which was there when Somerled was killed at Renfrew and which was in the blood spilt on the shores of the Clyde. And the chromosome which is still there, deep inside his bones buried somewhere beneath the thin soil of that windswept land.

I had managed to find this chromosome not just in the clan chiefs but in a great many other men bearing the name. It was impressive to find the same Y-chromosome in all five of the chiefs, but it was a surprise that so many other members of all three clans could also now claim to have a direct and unbroken line back to Somerled himself. Among the Macdonalds who volunteered their DNA, 18 per cent had inherited Somerled's Y-chromosome. The proportion among the MacDougalls was higher – 30 per cent of MacDougalls had his Y-chromosome in their blood – and higher still among the MacAlisters, almost 40 per cent of whom carried the clan founder's Y-chromosome. Admittedly, this was a relatively small sample, but why was there such a difference? Surely those bearing the name Macdonald should include more of Somerled's descendants than the other surname groups? Initially I was surprised that the MacDougalls and MacAlisters, who were, in my mind, somehow less directly connected to Somerled than the

Macdonalds, had actually inherited his Y-chromosome in greater proportion. But when I went over these results with Margaret Macdonald, the archivist at the Clan Donald centre, the explanation suddenly became clear.

The reason I had never expected to find any detectable association between Scottish clan names and Y-chromosomes was because of the widespread practice of name adoption, which I have already mentioned. I was pretty sure that this would drown out any authentic genetic signals from a common ancestor, like Somerled, because so many men would have taken the name of their clan chief, without being related to him. But the results speak for themselves. Against all the odds, there really is a clear and consistent Y-chromosome signal from the common ancestor himself, not just in the clan chiefs but in a great many others. But why is the proportion of men who have inherited Somerled's chromosome higher among the men of Clan Alastair and Clan Dougall than among those of Clan Donald? The answer, I believe, lies in the relative wealth of the three clans and the lands they controlled. Clan Donald is by far the biggest clan of the three. Through the acquisitions of their ancestors, starting with Somerled's son Ranald, the clan became by far the most important and influential in the west of Scotland. With so much land under Clan Donald control it is no surprise that so many men took the name. Clan Dugall, on the other hand, forfeited much of its land when it backed the losing side in the war between the English king Edward II and Robert the Bruce in the early fourteenth century, which culminated in victory for Bruce at Bannockburn. A smaller clan with less land and fewer people adopting the name would mean that a higher proportion of MacDougalls would be genetically related to the chief. And that is exactly what we found. Clan Alastair has always been the smallest of the three clans, with the least land and so with even fewer people having good reason to take the name, and among the MacAlisters an even greater proportion are related to the chief of the clan.

The real surprise is that so many men are directly descended, through an unbroken paternal lineage, from the founder of each clan and, further back, from Somerled himself. The numbers are astonishing. Take the Macdonalds. There are somewhere in the region of two million male Macdonalds worldwide. If the proportion sharing Somerled's chromosome in our sample is representative of all Macdonalds, and there is no reason I can think of why it should not be, then there are something like four hundred thousand men with Somerled's Y-chromosome alive today. Add in the MacAlisters and the MacDougalls and the number approaches half a million. That is half a million copies of a Y-chromosome made from just one original in the space of only nine hundred years. Had we stumbled across the world's most successful Y-chromosome?

Somerled's own traditional genealogy stretches back through his father, Gillebride, to his grandfather, Gilledomnan, and back to the kings of Ireland – to Colla Uais in the fourth century and as far back as the legendary Conn of the Hundred Battles in the second century. This is a fitting pedigree for a Celtic hero. However, I do not think it can be accurate, for the following reason. Somerled's Y-chromosome is a class 3 – a type that is almost unknown in Ireland outside the Scandinavian enclaves. From a study organized by Dan Bradley and his colleagues at Trinity College Dublin it is pretty clear that more or less all of the Irish Y-chromosomes that were around in the first millennium AD were in class 1. So Somerled's chromosome is in the wrong class to have come from the long line of Irish kings that is claimed for him in the traditional genealogy. It is also a rare chromosome in Scotland outside the three clans. But the one place it is not rare is Norway. We have found six exactly matching chromosomes, and many that are very closely related to it, among the samples from volunteers which Jayne and Eileen brought back from Oslo. This is a classic Norse Y-chromosome. On this evidence Somerled, the Celtic hero, was directly descended from a Viking.

Whatever its origin, Somerled's Y-chromosome has had a spectacular career since his death in 1164. In the space of less than a thousand years it has produced half a million copies of itself. This is sexual selection of a sort – and on a grand scale. What was it about this chromosome that made it so successful? Was it anything intrinsic to the Y-chromosome itself? I doubted that – it had spread far too quickly. Somerled's Y-chromosome had succeeded because it had benefited from the assets of wealth and status to which it had become inextricably linked and then from the patrilinear succession that kept these privileges closely tied to it in successive generations. I wondered if there were other Y-chromosomes to be found whose brilliant careers were launched by the same intoxicating sexually selected cocktail?

17

THE GREAT KHAN

While I was in Scotland unravelling and marvelling at the extraordinary success of Somerled's Y-chromosome, other researchers
had stumbled across an even more amazing Y-chromosome success
story. Tatiana Zerjal and Chris Tyler-Smith from Oxford had
noticed a similar irregularity in the evolutionary network of Y-
chromosomes that they were studying from Mongolia. Just as
Somerled's Y-chromosome stood out as an unusually large circle in
the Scottish network, so one particular Y-chromosome was far
more common than any of its neighbours on their evolutionary
networks. Just like Somerled's chromosome, this one became
apparent as a large central fingerprint with a few satellites
surrounding it, the signal of recent mutations branching off from a
founding chromosome. By counting these mutations and factoring
in the mutation rate, Tatiana and Chris established that the
common ancestor of this prolific Mongolian chromosome had
lived about a thousand years ago. They began to look for it in
other countries and, to their astonishment, found exactly the same
Y-chromosome dispersed across a great swathe of Asia stretching
from the Pacific in the east to the Caspian Sea in the west. What
explanation could there be for such a result? And then the penny

183

dropped. The range of this Y-chromosome corresponded precisely with the boundary of the Mongol empire founded by that most feared of all conquerors – Genghis Khan.

Genghis Khan was born around 1162, two years before the death of Somerled on the other side of the world, into the ruling family of a powerful local clan. Orphaned in his teens, he saw his family lose most of its power; but through skilful alliances and success in tribal wars, by the age of forty-four he was able to have himself proclaimed ruler of All the Mongols and took the title of Genghis Khan, or Great Leader, with a divine right to rule. After consolidating his grip on Mongolia from his capital Karakorum he embarked on a ferocious campaign of military conquest. His army, though not especially large, was well organized and disciplined, its superb horsemen and archers putting to deadly military effect the natural talents of a nomadic people who had herded and hunted on the vast prairies of their homeland for millennia. First he broke through the Great Wall and subdued the Chin empire of northern China. Then he led his army west and conquered parts of what are now southern Russia, Kazakhstan, Afghanistan and Iran. By the time of Genghis Khan's death in 1227, his empire stretched for five thousand miles from the China Sea in the east to the Persian Gulf in the west. The empire was divided by his principal wife among his four sons, who each continued and extended his conquests. His third son, Ogadei, who succeeded his father as the Great Khan, ruled from the eastern part of the empire, which by then incorporated Korea, Tibet and a large part of China as well as Mongolia itself. The rest of China was brought into the empire by Genghis's grandson, the great Kublai Khan, when he defeated the Sung dynasty. He moved the capital of the empire from Karakorum to Beijing, but failed in his ambitious attempts to conquer Japan and Java.

To the west another of Genghis's grandsons, Batu, began the invasion of Europe. In daring winter raids, when his cavalry could move quickly along the frozen rivers, Batu swept across northern

Russia in the only ever successful winter invasion of that country. He then destroyed Kiev, the capital of Ukraine, attacked Hungary and Poland, crushing a Christian army at Legnica, and even reached the Adriatic. Western Europe was saved from full-scale invasion only by the death of the Great Khan Ogadei in 1241, after which Batu withdrew to the eastern empire to contest the succession. The Mongols nevertheless kept their grip on their western empire and extended it to the banks of the Tigris, attacking and capturing Baghdad in 1258. At its height at the beginning of the fourteenth century, the Mongol empire was the greatest land empire the world has ever seen, before or since. By the end of the same century it was crumbling. Split by rivalries between the decendants of Genghis Khan and in conflict with three religions, Christianity, Islam and Buddhism, the great empire gradually fell apart, losing first southern China, to the Ming dynasty in 1367, then the western empire as it disintegrated into local khanates.

For all his fearsome reputation as a merciless and ruthless warrior, Genghis Khan was an unusual empire-builder. Though he frequently sacked cities and massacred their inhabitants, and was merciless with defeated armies, this was not done just out of pure savagery but as a necessary means to break the opponent's power. He also showed no interest in the cultivated urban pursuits of his conquered nations. He was clear from the outset that his Mongols should remain warrior nomads on the open steppes, merely using the cities and farms of their conquered lands as sources of revenue to fund their own ancient way of life. The way he went about his military campaigns was also ideal for the propagation of his Y-chromosome. According to one contemporary source, the plunder of a defeated enemy's lands could begin only when Genghis Khan gave permission; after that, all ranks had equal privileges – with one important exception: all the beautiful women had to be handed over to Genghis Khan himself. Even his doctor advised he slept alone 'from time to time'.

The present-day geographical distribution of the Y-chromosome in question, undoubtedly descended from one man in the last thousand years, fits so well with the limits of the Mongol empire at the time of Genghis Khan's death that it seems to me extremely likely that Tatiana and Chris have indeed found the chromosome of the man himself. What is truly amazing is the proportion of men living in these regions today who have inherited the Khan chromosome. In the sixteen different locations that were sampled, the chromosome is found, on average, in a staggering 8 per cent of all men. If this proportion holds for the entire region, that makes a total of 16 million men who now carry the Khan chromosome. This trumps the Somerled chromosome by more than thirty times and makes the founder of Clan Donald look like a very local Lothario.

But how sure can we be that this really is the Khan chromosome? While the identity of the Somerled Y-chromosome is not in doubt, thanks to the DNA match with the five living chiefs of Clan Donald, the same tests cannot be done on the Khan Y-chromosome. No-one knows where he is buried, nor are there any directly documented descendants. Though the circumstantial evidence for the Khan chromosome is strong, the proof is missing. However, there is one other piece of evidence that supports it. The Khan chromosome is practically unknown outside the limits of the Mongol empire – except in one place. Among the Hazaras, a tribe that lives on the borders of Afghanistan and Pakistan, the Khan chromosome reaches its highest frequency anywhere. Almost one-third of Hazara men carry the Khan chromosome, while in neighbouring tribes it is completely unknown. Through genealogies passed down as oral history many Hazaras claim direct descent from Genghis Khan himself. It is not proof, but oral histories have a habit of being proved right when genetics investigates.

The Khan chromosome has multiplied with amazing speed – one to sixteen million in about thirty generations. It has been provided with all the advantages, making its debut on the international stage in the loins of a sexually voracious and extremely successful military conqueror and being boosted in later generations by the rules of a patrilinear succession which bestowed on its later hosts the wealth and power required to continue the family tradition of sexual excess. This computes as a selective advantage of almost unheard-of proportions. It is also an entirely new type of evolutionary mechanism: a selective advantage for a Y-chromosome obtained through the very system triggered by the chromosome itself through its agent testosterone – aggression, conquest, promiscuity and patrilineal succession. This isn't sexual selection on the model of the peacock's tail, where the males compete and the females choose. The males compete all right, but the element of female choice is hard to see among the women lined up for insemination by the Great Khan after a battle. I would wager that no gene in human history has done as well as the Khan chromosome. So well has it performed that it is really hard to tell who is in charge. Is the Khan chromosome's achievement down to the sexual exploits and military conquests of the Mongol emperor? Or was the Great Khan himself driven to success in war, and in bed, by the ambition of his Y-chromosome?

Adam's Curse was becoming clearer. We had shown how Y-chromosomes had benefited from the seduction of native women in Polynesia, the Spanish conquest of South America and the violent raids of the Vikings. We had identified individuals with vast power and wealth obtained through violence and conquest. This is a new variety of sexual selection, based in part on female choice but also on female coercion. Y-chromosomes really don't care whether the eggs are willing or not.

18

THE OLD SCHOOL REGISTER

The research that led to the discovery of Somerled's Y-chromosome was very exciting – and very rewarding. It had led, completely unexpectedly, to new information of genuine historical interest. This is the genetics I prefer – the genetics of real people, real ancestors. It is alive. The identification of Somerled's Y-chromosome, and also of Genghis Khan's, came about through something that didn't fit. The smooth transitions between a Y-chromosome and its mutational derivatives on an evolutionary network, where the descendant chromosomes diverge slowly away from the abundant original, were just not there. That was the pattern I was expecting from my experience with mitochondrial DNA. Instead of that, some Y-chromosomes, represented by circles on the interwoven evolutionary network, were much more abundant than they should have been and others were much rarer. Sometimes, there was no sign of Y-chromosomes where I would have expected to see them. In my years of putting together mitochondrial networks I had never seen anything like this. There were virtually no empty nodes, as we called them, in mitochondrial networks – hardly any intermediate mDNA sequences that should have been detected in at least some individuals but were never found.

This was the evolution of a gene of continuity.

Not so with the Y-chromosomes. There were all sorts of irregularities: empty nodes, tiny circles on their own at the ends of long filaments, big circles next to little circles. Whatever was going on with the Y-chromosome, it was very different from mitochondrial DNA. It was as if individual Y-chromosomes had suddenly exploded into life, multiplying furiously with no regard at all for their theoretical obligations. By lucky chance we had spotted one of these eruptions on our charts and, by even greater good fortune, linked it to a historical figure. That one Y-chromosome of Somerled, virtually on its own in 1100, had by the year 2000 increased in number by half a million times. How had it managed this amazing feat? The general increase in the population of Scotland doesn't even begin to account for it, even when you take account of the number of Scots who emigrated and their descendants. If a Y-chromosome had just kept pace with the general increase in the population it might have gone from one in 1100 to perhaps twenty, fifty or even a hundred in the present day. It's hard to be precise, but there is no need to be. You don't need statistics to tell you that it is nowhere near five hundred thousand. We were not looking at a plodding improvement over the centuries. This was a supernova. How had it managed it, and how had Genghis Khan eclipsed even Somerled's magnificent genetic achievement?

I already knew the answer. It lay in the story of Somerled himself – the story which I have given you in such detail. Somerled was powerful. He was wealthy. He had land. He passed this wealth on to his sons, and they and their descendants in turn became the chiefs of mighty clans. If Somerled had lost the sea battle against Godred of Man off the coast of Islay on that dark winter's night, his Y-chromosome would be invisible among a million others. I looked again at the network, at the empty nodes: Y-chromosomes that must once have existed but were no longer there – or, if they were, that we had never found. Were these gaps left by the

Y-chromosomes of ancestors who had lost their battles, who had not acquired wealth and had nothing to pass on to their sons? Were the fuller circles on the network, the Y-chromosomes that were much more common than they should be, the genetic legacies of the material success of their ancestors? The network slowly became a history of success and failure; of Y-chromosomes diminished or extinguished by some mischance and of others that had flourished. Was this the real message filtering down from Somerled through the swirling mists and howling gales of the Isles?

Very gradually, vague images began to form in my mind. Was Somerled's message coming from the man or from his Y-chromosome? Was he the architect of his chromosome's success? Or was he the instrument it used to propagate itself? The more I thought about it, the more I felt the whole scene reversing. It was as if the stage of history was being turned around and I could see behind the scenery to the puppeteers who pulled the strings. They had become transformed into the chromosomes I had seen down the microscope, but instead of being fixed to a glass slide, they were oscillating like some strange larvae. And at their centre, like a bloated maggot, more active than all the others, was the pale form of the Y-chromosome itself. It had no eyes and the frantic writhing of its pallid, segmented body disrupted the choreography of the other chromosomes as they tried in vain to work the strings. The stage kept turning and when it had gone full circle the savage and dislocated play of life made sense. The longships casting off into dark Atlantic waves, the cries of the murdered monks of Lindisfarne, the slaughter on the shores of Morvern, the thunder of Mongol cavalry along frozen Russian rivers, the blood of defeated enemies and the screams of their women as they were led away to the Great Khan – all these were caused by the blind squirming of the Y-chromosome as it writhed behind the scenery. The image faded, but I have never forgotten it.

I looked again at the networks. Were these explosions and

extinctions explained by wealth, conquest and power, the indirect manipulators of sexual selection, or by an intrinsic quality of particular Y-chromosomes? Somerled's and Genghis Khan's Y-chromosomes had proliferated so wildly because of their power, but could there be another, additional reason why their chromosomes had persisted right down to the present day? I remembered that William Hamilton had once predicted that any Y-chromosome which mutated to produce just males would spread very quickly. Had we stumbled across real-life examples of Hamilton's theoretical superselfish Y-chromosome? It was difficult to disentangle the possibility from the wealth and power that also accompanied the chromosome as it snaked through the generations. I then began to wonder about my own Y-chromosome. It had definitely increased way above theoretical expectations, from just one in 1300 to about ten thousand now – a performance not nearly as impressive as Somerled's but way above what chance alone would predict. The Sykes Y-chromosome had done very well, and without ever being rich and famous.

There were, it seemed to me, two possibilities to explain the unusual pattern seen in the evolutionary networks of Y-chromosomes. In exceptional cases, like Somerled and Genghis Khan, there had to be a degree of sexual selection going on even to begin to account for the astonishing proliferation of their Y-chromosomes. It doesn't take a genius to realize where this sexual advantage lay: in wealth, status and power. The continuing prosperity of these Y-chromosomes over the centuries had been sustained by the inheritance of just those things, thanks to the rules of patrilinear transfer which made sure that wealth and status, badged in Somerled's descendants by the accompanying surname of the dominant clan, Macdonald, followed the same course as the chromosome through the generations. But in other cases, including my own ancestors, where there was (as far as I know) no wealth, power or status to act as sexual attractor, I wondered if the Sykes Y-chromosome had done well because of an intrinsic ability to have more sons than daughters. Was it a mini-version of a Hamiltonian superselfish Y-chromosome, making good because of some inbuilt quality rather than being propelled by its association with wealth and property? A peasant chromosome it may be, rooted in the chilly hillsides of Yorkshire, but it could dream as well.

Had my own surname proliferated by some means other than by pure chance? The standard version of human sex determination, which I myself had never felt any reason to doubt and which I had always taught my students, is that each pregnancy has an equal chance of being male or female. Since sperm containing X- and Y-chromosomes are produced in equal amounts, an egg might just as easily be fertilized by a sperm containing an X-chromosome as by one that carries a Y-chromosome. But what force lies behind the easy calculus that equates the sex of a child with the tossing of a coin? If some surnames were common because their Y-chromosomes were somehow managing to get themselves over-represented in each generation, this would be amazing. If something about a Y-chromosome were consistently able to distort the sex ratio in its favour by only a small amount, then its career as a chromosome would be infinitely brighter. If Sykeses produced even 10 per cent more sons than daughters at each generation, this seemingly small advantage would go a long way to explaining how the name (and the chromosome) had increased from a frequency of one in the thirteenth century to over ten thousand today.

It had always been assumed that surnames came and went by a process of random chance. Surnames would vanish when the last male either had no children at all or, more commonly, had only daughters. The name, in the phrase of genealogists everywhere, would have 'daughtered out'. If the birth of boys or girls were determined entirely randomly, then so was the fate of a surname.

I had raised this question with George Redmonds, the Yorkshire surname expert, when we were walking along the winding stream near the village of Flockton in our search for the home of the original Mr Sykes. It turned out he had been wondering why some surnames became common while others remained rare and even vanished altogether. When I asked him if he thought an explanation for the rise of some surnames and the decline of others might be that some families had more sons than daughters, he agreed it was certainly a possibility, though without experiencing the same thrill of heresy that I began to feel as a geneticist.

It was certainly true that a few surnames had come to dominate the Colne valley in Yorkshire where the Sykeses were concentrated, and the same is true of any country district. I remembered when I did the post round during the Christmas holidays near my parents' home on the borders of Suffolk how two surnames, Ablitt and Mathews, must have made up a good third of the deliveries. I had occasionally wondered about that during my years of teaching genetics and, rather lazily, put it down to the random chance of having a son or a daughter. That process, called genetic drift, is a powerful one in small communities and very soon eliminates most of the surnames without recourse to any other more exotic mechanism, such as the one that was brewing at the back of my mind. To persuade you of the power of genetic drift, let us imagine we are back in the thirteenth century at the period when English peasants are being given their names.

We are in the imaginary Yorkshire village of Flockthwaite, where live eight couples. Their newly acquired names are Bubblefroth, Winkleweed, Redbelly, Oakenthigh, Jackersnipe, Silverspoon, Barraclough and Sykes. Each has two children. Purely by chance the Bubblefroths and Winkleweeds have two daughters each. That's the end for these two names. The Redbellys, Oakenthighs, Jackersnipes and Silverspoons each have a boy and a girl. But the Barracloughs and the Sykeses each have two boys. In a single generation two surnames have daughtered out. Now there is one male each of Redbelly, Oakenthigh, Jackersnipe and Silverspoon but two Barraclough boys and two Sykes lads. They all marry and have two children each. This time the Redbellys and Oakenthighs have two boys, the Jackersnipes and Silverspoons have two girls and the Barracloughs and Sykeses all have a boy and a girl. No more Jackersnipes and Silverspoons. In just two generations the population of Flockthwaite still has only eight couples but we have already lost four surnames. The Redbellys, Oakenthighs, Barracloughs and Sykeses are still battling it out. Pretty soon they will disappear one by one as they daughter out until there are only two surnames left. They will vie with each other for a few more generations until one vanishes and everyone ends up with the same surname. For a small village the size of Flockthwaite, with only eight couples and a static population, this process takes, on average, eight generations to get down from the original eight surnames to just one. If the population of Flockthwaite grows over the years it means couples are having more than two children, so it takes longer for surnames to daughter out simply because everyone has a better chance of producing a son. But eventually it will still happen.

Now let us imagine that one of the Y-chromosomes in Flockthwaite, attached to one of these names, has worked out a way of having more sons than daughters. To go to the extreme, we will suppose that one name only ever produces sons. I put the eight names into a hat and drew out Oakenthigh as the favoured name/chromosome. The other couples carry on as before. Two names daughter out in the first generation, another two in the second generation. Now there are four Oakenthighs and one male each of three other names. By the third generation, there are eight Oakenthighs and the other names have almost gone. I have had to cull some of the sons to avoid the population of Flockthwaite increasing and I have also had to import some females, but the overall effect is dramatic. In all the simulations, Oakenthigh always ends up being the sole surviving surname, and it does so very fast – on average, in only four generations.

This shows how handsomely it pays a Y-chromosome to be able to produce only sons. In my extreme example Oakenthigh always ends up as the dominant name in Flockthwaite. But even a more modest tendency to produce sons greatly increases the chance of a surname being the sole survivor in a community, though it might take a little longer. Although the completely random process we looked at first will indeed whittle down the eight surnames to just one in time, each surname stands as good a chance as any other of winning the race to be top dog in Flockthwaite. But suppose a surname did have a tendency to produce more sons than daughters; that would certainly help a lot. But does it happen? Does this explain why some surnames are very common in a locality? No-one seems to know. The existence of a powerful random mechanism to explain the evidence for abundant surname survival, and extinction, may have meant that those people who think about such things had paid little attention to the possibility. However, since hardly anybody had ever thought that so many names had single genetic founders, the extraordinary success of some names was not properly appreciated.

The research with my own name, by contradicting the received wisdom, had brought this question into focus. There had been only one founder – or, if there had been others, they had not done very well. Only the Y-chromosome of Henri del Sike had prospered. Unlike the case of Somerled and the Macdonalds, there was no reason I could think of why anyone would have wanted to adopt the name Sykes; the Sykeses were never wealthy or powerful. And Sykes wasn't the only name that showed this remarkable association with a single original Y-chromosome. Could it be that there was something special about these Y-chromosomes? Did they produce more sons than daughters? I asked George if there were any other names in the vicinity that might also have single founders, even though none had been suspected. He suggested Dyson. This is another Yorkshire name and it had been assumed for ages that, rather like Sykes, it had multiple independent founders. Whereas Sykes comes from a common feature of the landscape, the name Dyson suggested an occupation – the son of a dyer. Medieval Yorkshire was full of dyers working in the wool trade and most Dysons assume that they had inherited the name because one of their ancestors had originally been the son of a dyer. And with hundreds of dyers around at the right time, there was no reason to suppose that only one of them had given rise to the name. Like Sykes, most people thought Dyson was common because there were a lot of different originals to begin with.

George, on the other hand, had a different idea. In his research in the court and estate records he came across a reference to a remarkable lady called Dionissia of Linthwaite. She was, by all accounts, a complete tearaway. More than once her name appeared with convictions for cattle rustling and other crimes. It was also recorded that in 1316 she had a son called John, though there is no mention of the father. The boy's surname was recorded as Dyson not because he was the son of a dyer but because he was the son of Dionissia, conveniently abbreviated to Di. This is an example of a very rare phenomenon, a matronymic rather than a patronymic surname. If George was right in his suggestion that the John Dyson born in 1316 really was the single founder of all Dysons living today then we should be able to pick up the signal of this common ancestry among the Y-chromosomes of modern Dysons. If, on the other hand, Dysons were originally the sons of several different dyers then we might expect a mixture of Y-chromosome signatures among modern Dysons.

When we got the results, they exceeded even our most optimistic expectations, eclipsing even the amazing outcome of the Sykes study. Of the twenty-three Dyson volunteers who sent us their DNA, nine had exactly the same Y-chromosome signature and a further eleven had chromosomes that were very closely related to it. Of the three Dysons whose chromosomes did not match the common one, two were very close to each other, and one was on its own, unlike any of the others. This was astonishing. Nearly 90 per cent of the Dysons had the same or related Y-chromosomes. George was right. There was only one founder. We had yet another name which had proliferated from just one originator. There are about five thousand Dysons living today – including the famous James Dyson, inventor of the bagless vacuum cleaner – who have inherited the name and the Y-chromosome from a single man. Had Dysons, like Sykeses, proliferated because of an inherited tendency of their Y-chromosomes to produce more sons than daughters?

What was equally astonishing about the Dysons was the extremely low non-paternity rate, arising from adoption and infidelity. We had found an indication of only two such events, the ones that dislocated the two separate Dyson branches from the rest. These could even have been descended from separate founders. Whatever the explanation for those disruptions, the incidence of non-paternity among the Dysons was extremely low over the seven hundred years since Dionissia the cattle rustler had her first son. It made the generations of Mrs Sykeses look like serial adulterers compared to the saintly succession of the Dyson wives. George Redmonds was absolutely delighted with the news. We had proved one part of his theory about the Dyson name, that they were pretty well all descended from one person. But could we prove this founder was John, son of Dionissia? Unlike the Clan Donald, there was no traditional genealogy to go on. There was no wealth, land or title involved in the Dyson succession, so nobody had bothered much about keeping records. Nor did we have the advantage with the Dysons, as we did with the Clan Donald chiefs, of being able to test the DNA of living descendants who could claim direct descent from the founder through the records.

Nevertheless, we could make a stab at the time that the original Mr Dyson lived by seeing how many mutations had occurred among his descendants. The rate at which genetic signatures change, the mutation rate, is not well known, and I am quite sure some of the DNA repeat elements of the signatures mutate more quickly than others. But people have used an average of one mutation in every fifty generations for the ten-element genetic fingerprint which we have adopted. On that very rough basis, we can work out how long it has taken for eleven out of the twenty very closely related Y-chromosomes to have changed by one mutation. We can get at the time by multiplying the proportion of these mutations (11/20) by the mutation rate (1 change per 50 generations) and, when we do the calculation, the answer comes out at a figure of 27.5 generations. It is hard to know what to take as the average generation time over the last 700 years, but if we use 25 years per generation, which is not so unreasonable, 27.5 generations have passed in 687 years. Since I did the calculation in 2001, this takes us back to 1314. John Dyson was born in 1316. Uncannily close. And, of course, misleadingly so. Had we increased the generation time in the calculation from 25 to 30 years, it would take us back to 1176. But the time was sufficiently close to add even more weight to George's theory about the identity of the first Dyson.

That was all icing on the cake as far as I was concerned. The important fact was that the Sykeses were not alone: several Y-chromosomes had proliferated way beyond expectations based on random genetic drift. Now I wanted to see whether the reason for this might be that the bearers of some names really produced more sons. Where could I look for the evidence? I had started my surname research with my own name and I thought I might as well continue with it. The best thing about it is that I got to meet a lot of relatives I never knew I had. Also, I didn't feel impertinent in asking questions about the Sykeses, as I would have had I been digging into the Dysons or some other family.

I began to ask the Sykeses I met while recording a radio programme with George Redmonds on genes and genealogy. Did they think there were more boys than girls born in Sykes families? It's very easy to get lost and confused while listening to people telling you about relatives you have never met. The answers to my question usually went something like this: 'Well, Hilton Sykes down at Slaithwaite Hall had four sons and Michael, when he moved to Ainsley Place, had two sons and a daughter. And those Sykeses down the valley had three boys, or was it two? And my grandmother was a Sykes, but, wait a minute, she had two sisters.' I enjoy talking to these folk, my distant relatives. Living, as I do, in a town, I am amazed and impressed, even a little envious, at how they all know each other. But, aside from reconstructing the entire pedigrees of Sykeses going back generations, was there a quicker way of getting to the truth? The stories I had been told about the Sykeses and their huge numbers of sons all come from around Slaithwaite in the Colne valley. I wondered if there was a way of finding out how many children there were in and around Slaithwaite called Sykes, and whether they were boys or girls, without having to rely on the vagaries of selective memory and what geneticists call biased ascertainment – which is a long way of saying 'finding what you are looking for'. The electoral roll would be no good because Sykes girls would be very likely to change their names on marriage and, conversely, women called Sykes may or may not have been born with the name. Then it struck me that, of course, all children have to go to school. Perhaps the school in Slaithwaite would have the records which would let me discover whether there really were more Sykes boys than girls.

Mary Pontefract is the administrator for Slaithwaite Church of England Primary and Infants School, which takes all the children from the town and thereabouts. She told me when I rang that she was sure she could lay her hands on the old admission records for the school and would be very pleased to let me have a look through them. A week later, I was on my way north. The steep-sided Colne valley, in which Slaithwaite lies, was green with spring grass when I arrived in the early evening. I was staying on a farm, in a building converted from a weaver's cottage. My room on the upper storey still had the four large windows along one side which admitted sunlight to the weaving loom. There are scores of cottages like this on the hillsides around Slaithwaite, a reminder of the days when each family would have a loom and make up pieces of cloth for sale. When the steam-powered mills opened on the valley floors, the household looms were put out of business and the people left their cottages to work in the grim towns that sprang up.

The journey from the farm to Slaithwaite school took me from the windy uplands the short distance to the town, down past sullen yellow-grey terraces pressed into the steep hillside and beneath the soaring arches of the stone viaduct which carries the Huddersfield to Manchester railway line. Mrs Pontefract was there at the school entrance to welcome me and she settled me into a vacant study. I was absolutely amazed to find she had unearthed the school registers going back over a hundred years, which she brought to me in a cardboard box. I took out the first of the registers and opened it. There was the faintest whiff of camphor, not strong but there in the background, and a musty smell of pure age and ancient collections. Strange how smells can sometimes bring back long-forgotten memories. In an instant I was back in a taxidermist's shop in the Strand in London – the shop, long gone, where my father used to take me after work; the shop where I had first seen collections of butterflies and where I bought the very first thing I ever owned – a very bald tiger's head, price £2. That shop had exactly the same ancient smell as drifted up from the register. The first page read: 'The Crown Register of Admissions, Progress and Withdrawals by J S Horn. Price 5 shillings. Delivered to J Quinn, Head Teacher, by H H Rose, correspondent on the Ninth Day of June 1893.' This book was over a hundred years old. By the look of the dozen or so other registers, from this collection I would be able to get a complete list of all the Sykeses from the late nineteenth century through to the present day.

This was not just a list; it was a document which, from its yellow pages and blue lines, was whispering to me the history of this little town. First of all, each entry was written in the most beautiful script. I could imagine a quill pen dipping into a bottle of deep black ink and slowly forming the letters of each child's name, when they were admitted and when they left. On the far right-hand column of the double-page entries, the destinations of the departing children brought alive both the certainty and the hopelessness of life in late nineteenth-century Yorkshire. The vast majority of both boys and girls had their entries closed in this final column by the single word 'Woollen' or 'Cotton'. After a few brief years at school, their life was the mill: one of the great six-storey blocks that spun and wove textiles for the world. Many mills still stand in Slaithwaite, a few still in production even now. But the main industry collapsed a long time ago, leaving the town today slightly shabby and uncertain.

But, as I turn the pages of the register, this is still a hundred years into the future. On these pages the great mills, like living, breathing animals full of energy and industry, issuing steam and toil in equal proportions, consume the children and dominate the lives of everyone. The final column of the register showed that those few who were not immediately bound for the mill took up other trades. Boys became clerks, rug-makers, cabinet-makers and errand boys. Girls who escaped the mill became, in the single-word summary of the register, 'domestic' or occasionally 'dressmaker'. Everyone found a job and the teachers cared enough to write down what it was. In later decades, the entries in this last column became more erratic until, in the register for the final decade of the twentieth century, they were replaced entirely by something called a UPN – doubtless an abbreviation for Universal Pupil Number or some similar bureaucratic invention which obliterates any picture of the child or the town.

There were over a thousand entries in the first Crown Register, and they could not have been more conveniently organized for my purpose. As well as individual entries for each child written in chronological order of their admission to the school, there were separate sections containing an alphabetical listing for each few years. I turned to the pages containing the surnames beginning with S. There were several columns of names and immediately I could see that, among them, were Sykes after Sykes after Sykes. It was by far the dominant surname beginning with the letter 'S'. At first glance it looked as though at least half of them were called Sykes. Which were boys and which were girls? Their first names made that distinction easy. Harry, George, William, Frank for the boys; Edith, Annie, Emily, Mary for the girls. Names today that are rarely used as one fashion replaces another. Often two children with the same name would arrive on the same day, and so were differentiated by variants: George and Georgie, William and Willie, Elizabeth and Lizzie. The school was crammed full of Sykeses. Other names, too, occurred with obvious regularity – Bamforth, Hirst, Dyson, Sutcliffe, Wood – each consuming row after row of neatly written entries. I checked each Sykes entry in the alphabetical section against the birth date in the main entry to make sure George and Georgie and other similar differences did, in fact, represent separate children, and then added up the boys and the girls. Of the seventy-six Sykeses – almost exactly half the S entries spanning the years 1886–94 – forty-two were boys and thirty-four girls. That was almost 25 per cent more boys than girls. If Sykeses really did have that many more boys than girls it would be a fantastic advantage to their Y-chromosome and go a long way to explaining why the name had become so common. It was a great start, but was this excess of boys going to be consistent?

For the rest of the day I went through all the registers from 1886 to the end of the twentieth century. It took a long time to make sure I had not counted any child more than once. Many had entries in consecutive registers and some children had been admitted and re-admitted several times, having been withdrawn for reasons that were unrecorded. As the nineteenth century passed into the twentieth, the number of Sykeses went up and up. Sad entries in the last column recorded the deaths of children while still at school. This tragedy, almost unknown nowadays, was much commoner in those days when infectious diseases – typhus, smallpox, tuberculosis and scarlet fever – haunted the crowded cities, unrestrained by antibiotics. In the register, first names began to change. Eric, Norman, Raymond, Margaret, Eileen, Amy in the 1920s. David, John, Keith, Pauline, Vivienne, Susan in the 1940s. I found Sir Richard's entry: pupil number 45, Richard Sykes, 7, Brookside, Slaithwaite; admitted 23 August 1948. Names changed again, to Mark, Karl, Wayne, Kimberly, Katie and Victoria in the 1980s. But in every register there were always slightly more boys than girls.

Out of a total of 393 Sykes children from Benjamin (b. 24 June 1860) to Timothy (b. 23 March 1989), 212 were boys and 181 were girls. The trend I had detected in the first register had continued. Over the course of a hundred years there were 17 per cent more boys than there were girls. That is almost five boys for every four girls. That may not sound a lot, but repeated generation after generation it would certainly, other things being equal, have a major impact on the survival of the surname. It really looked as though the hunch I had discussed with George Redmonds, and the folklore of the valleys, had some substance after all. There really had been more Sykes boys than Sykes girls. Could there have been any influences that I had overlooked? For example, was it possible that more boys than girls attended the school? I had thought that was unlikely when deciding to search the registers in the first place and, when I asked her, Mrs Pontefract could think of no reason why it should be so. All children had by law to attend school throughout the entire period covered by the registers. And, as Mrs Pontefract pointed out, if any children had been sent elsewhere for a private education it was far more likely to have been the boys in a household than the girls. That would have reduced rather than increased the number of boys on the register of this state-run school. No-one I have spoken to since can think of a reason why there would be a bias towards boys in the district attending the school.

An equally important question is whether the difference between the numbers of boys and girls is a significant one. By that I mean, could the fact that we have more boys than girls attending the school be the result of sheer chance rather than being a consistent feature of the Sykes name? Could the total of 393 Sykes children, of whom 212 were boys and 181 were girls, have been generated by the same random process of sex determination that I had set out to investigate – the random chance of an egg being fertilized by a sperm with an X- or a Y-chromosome? One way of looking at this is to ask how often you might expect this result from a completely random process, like tossing a coin. What is the chance of throwing 212 heads (sons) and 181 tails (daughters) from 393 attempts? Without boring you with the details of the calculation, I can tell you that it comes out at just under 6 per cent. In other words, in 94 per cent of attempts the numbers of heads and tails would be closer to the expected 1:1 ratio than the 212:181 boy:girl ratio I found at Slaithwaite. If you are still with me, it means that there is a possibility (6 per cent) that the Slaithwaite results are a statistical fluke, but an almost sixteen times greater possibility (94 per cent) that they are not. I readily admit that this does not amount to an overwhelming probability, and I do not want to exaggerate the importance of the results from Slaithwaite. It indicates a certain ambition among Sykes chromosomes to rise above their station, but it does not prove it. I began to look elsewhere for more evidence for superselfish chromosomes.

19

THE ELEVEN DAUGHTERS OF TRACY LEWIS

Is there a tendency for certain families to produce more sons than daughters? I was soon to discover that musing on the equality or otherwise of the relative numbers of girls and boys born has been going on for centuries. The first scientific paper that I have found on the matter was published in 1710 in the world's first proper scientific journal – *The Philosophical Transactions of the Royal Society*. There are many Royal Societies of this and that, but only one Royal Society cited with no qualifying suffix (though its full title was the Royal Society of London for Improving Natural Knowledge). It was founded in 1660 and gained its Royal Charter two years later from Charles II ; today, being elected as a Fellow of the Royal Society is – short of winning a Nobel Prize – the pinnacle of achievement for a scientist.

Although I could read the paper in the facsimile editions held in the Bodleian Library in Oxford, I wanted to feel and smell the original volume if I possibly could. I had found the Slaithwaite school records so much more rewarding in original form than copies could ever be. Assuming the Royal Society library would have an original volume, I called in to their headquarters in an elegant Regency terrace a stone's throw from St James's Park in central London. I climbed a flight of marble steps past busts and portraits of Fellows, then past a great wall plaque engraved with the names of the past Presidents of the Royal Society. It read like a history of British scientific achievement. Joseph Banks the botanist, who accompanied Captain Cook on his first expedition to the South Pacific; Humphry Davy, physicist and inventor of the miner's safety lamp; Lord Lister, pioneer of antiseptic surgery; Lord Rutherford, nuclear physicist and discoverer of the alpha particle. Having explained my purpose, I was soon settled at a desk with the leatherbound original – now almost three hundred years old. The title page summed up the curiosity of the times.

Philosophical Transactions.
Giving some account of the Considerable Undertakings,
Studies and Labour of the Ingenious
in many Considerable Parts of the World.

Very gently, I turned the faded cream pages – pages that were saturated by the atmosphere of the great library and, for all I knew, contained molecules of famous scientists trapped within their woven fibres. The pages quivered and crackled as I turned them over, passing an account of the eclipse of the moon on 1 February 1701, a paper on the usefulness of the silk of spiders and a very long description of the bones of an elephant 'which died near Dundee on 27th April 1706'. On page 186 I reached what I had come to see.

An argument for Divine Providence, taken from the constant Regularity observed in the Births of both sexes. By Dr John Arbuthnott, Physician in Ordinary to Her Majesty, and Fellow of the College of Physicians and the Royal Society.

As Queen Anne's doctor, Arbuthnott would have been a busy man. The Queen was almost permanently unwell. Indeed, the front page of this very volume celebrated her return to good health – a return which, sadly, was only short-lived. Despite Dr Arbuthnott's undoubted skills, she died four years later, aged forty-nine, without leaving an heir. This was both remarkable and tragic because by the time she was thirty-five Anne had been pregnant at least seventeen times. Many of these pregnancies miscarried and not one of her children survived beyond childhood. I don't know whether or not it was this sequence of grievous losses afflicting his principal patient which concentrated Arbuthnott's mind on the peculiarities of childbirth, but he was sufficiently interested to plunge into the records to retrieve the numbers of boys and girls christened in London for the previous eighty years. In those days there was no compulsory registration of births and many infants would have died before they were first officially recorded at their christening. From his list, it was obvious that more boys were being christened than girls, so presumably more were being born as well. For every one of the eighty years there were consistently more boys, but what intrigued Arbuthnott was the regularity of the proportions. He was living at a time when calculation was extremely tedious, and he does not actually work out the ratio between boys and girls born for each year. With a modern calculator it took me only a few minutes to discover what would have taken him hours of long division to achieve. But he is absolutely right. The ratio of boys to girls is remarkably consistent over that period. It averaged just over 1.06, very close to today's value, and varied between extremes of 1.01 in 1703 and 1.15 in 1661. As well as being handicapped by the lack of a calculator, Arbuthnott was working at a time way before the development of statistics and two hundred years before there were even any hints of the genetic principles which actually decide a child's sex. To Arbuthnott, the consistency could not possibly be attributed to chance. He does some maths to make his point; then he uses it as an example of God's design, with the following conclusion.

Among innumerable Footsteps of Divine Providence to be found in the Works of Nature, there is a very remarkable one to be observed in the exact balance that is maintained, between the number of Men and Women; for by this means it is provided that the Species may never fail, nor perish, since every Male may have its Female, and of a proportionable Age. This Equality of Males and Females is not the Effect of Chance but Divine Providence, working for a good End, which I thus demonstrate.

His thought on the consistent excess of boys makes interesting reading:

we must observe that the external Accidents to which are Males subject (who must seek their Food with danger) do make a great havock of them, and that this loss exceeds far that of the other Sex, occasioned by Diseases incident to it, as Experience convinces us. To repair that Loss, provident Nature, by the Disposal of its wise creator, brings forth more Males than Females, and that in almost a constant proportion.

And here he makes a moral point – one which Genghis Khan would not have appreciated:

From hence it follows, that Polygamy is contrary to the Law of Nature and Justice, and to the Propagation of the Human Race; for where Males and Females are in equal number, if one Man takes Twenty Wives, Nineteen Men must live in Celibacy, which is repugnant to the Design of Nature; nor is [it] probable that Twenty Women will be so well impregnated by one Man as by Twenty.

The regularity of the sex ratio which Arbuthnott was the first to record, at least in modern times, and the consistent excess of males, was noted time and again over the next three centuries. Roughly 5–6 per cent more boys than girls are born nowadays just as they were in the early eighteenth century. Arbuthnott put this down to Divine Providence, but most subsequent commentators have interpreted it as a way of compensating for the greater infant mortality among the more sickly boys so that, after puberty when it is time to breed, the numbers even out. There is a whiff of the 'good for the species' about this; in fact, Arbuthnott himself makes just that point in adducing evidence of God's guiding hand, though he puts higher male mortality down to their working too hard as adults.

One of the later scientists to dabble in the human sex ratio was the geneticist R. A. Fisher, whom we met briefly in chapter 9. Fisher's aversion to group selection and his precocious focus on genes led him to explain the balance between the sexes at the critical period – when they are breeding – as an equilibrium between two opposing genetic influences. He thought that there must be genes around which would tend to make parents produce more sons and others with the opposite effect, so that the overall ratio of boys to girls was kept constant. This was a purely theoretical speculation on his part, an explanation for the constancy in the ratios of boys and girls born that had so impressed Dr Arbuthnott. Even so, if these genes existed, then maybe their influence would be found in the balance between boys and girls in different families. But are there families around in which there really is a tendency to have children of one sex or the other?

This is precisely the situation in which it is very easy to be misled. We all know families where the children are all boys or all girls, couples who keep on having children of the same sex in the understandable but often unrealized ambition of having one of the other sex. We notice these families more than those with a mixture of both sexes, and I guess most of us have a feeling that there is something other than pure chance that is deciding the sex of the baby in these cases. But human intuition is notoriously unreliable when it comes to assessing whether events are happening randomly or not. We are all inclined to see patterns where there are none, whether in the roulette wheel or the lottery. It is no different when it comes to the sex of babies. We might know in our rational mind that nothing but chance decides the number of the next lottery ball to be chosen and that no system on earth can predict which one it will be, yet still we believe we see patterns in the sequence of the balls.

Only a few days after I had been in the marble confines of the Royal Society, I was rung up by a friend who, knowing my interest in the topic, alerted me to an article in a celebrity magazine. I bought the magazine and there, on a double-page spread under the heading 'Baby Fever', was a photograph of the Lewis family from Dorset. It began: 'Tracy Lewis is addicted to having babies, but her husband doesn't seem to mind – which is why they've got another one on the way!' Tracy Lewis was pregnant again – for the thirteenth time – and the photograph showed her with her husband, Peter, and her twelve children in descending order of age: Carly (19), Tracy (17), Samantha (16), Charles (15), Lyndsay (14), Danielle (12), Chantelle (10), Charlotte (9), Georgia (8), Candice (6), Shannon (3) and Shaznay (2). What the article had not mentioned, and what my friend had noticed straight away, was that eleven of the twelve children were girls! Was this just a co-incidence or was there something else going on in the Lewis family?

I arranged to go and see them and one day in late December arrived at their house, which was festooned with coloured Christmas lights, in a neat suburb of Bournemouth. Inside, there were girls everywhere: on the sofa, on the floor watching television, in the kitchen; the youngest was still being carried by her mother, Tracy. The family were getting used to their celebrity, having been twice on television already with another appearance due the following month. They had endured many visits by journalists who naturally wanted to know how they managed with such a large family, what the only boy, Charles, thought about being brought up with so many girls and how they were all looking forward to the birth of the next baby. They hadn't had a visit from a genetics professor before, and I wasn't there to enquire about their domestic arrangements, but I couldn't help being fascinated by this delightful family – and by little details, like having two washing machines on the go all the time, cooking turkeys instead of chickens because a chicken is far too small for fourteen hungry mouths, that sort of thing. What I was intrigued to discover was whether this was a family which contradicted the rule that the chances of having a boy or a girl at each pregnancy are more or less equal. I had already calculated the odds of having a family with eleven of one sex and one of the other as three in a thousand. In terms of flipping coins it means if you tossed a coin twelve times over and over again, one thousand times, you would expect to come up with eleven heads and one tail on only three occasions. Which is not very often. So that might lead you to think there was something going on which was biasing the coin. But could it be just another example of biased ascertainment, of noticing the extremes and not the many other families with twelve children but with a more even split of sexes among them.

I was intrigued to know whether this tendency, if indeed there was one, ran in the family. Certainly Tracy and Pete fully expected their thirteenth child to be a girl. In fact, they told me that they were astonished that, after only three girls, their fourth child, Charles, was a boy. But what about Pete's, and particularly Tracy's, own family? Had they also been surrounded by sisters and not brothers? Over tea and biscuits, Pete brought out the details of both their family histories which he and Tracy had prepared ready for my visit. I began to draw the family tree starting with Pete. He had two brothers and two sisters, so nothing unusual there, and in his parents' generation, three aunts and two uncles – again, nothing out of the ordinary. However, when we came to Tracy's side of the family, there were far more girls than boys. Tracy herself had five sisters and one brother; her mother was one of three girls, though with one brother, as was her grandmother. Counting up the girls and boys among her children, siblings and maternal relatives, there were twenty-three girls and four boys. The chances of that happening, if the expectation for each child was 50:50, is one in five thousand. These are very long odds indeed; but, even so, it could still be a chance result. I had deliberately sought out this family. It is hard *not* to think there is something other than pure chance deciding the sex of the Lewis children, and we will see what that might be later. But are the Lewises exceptional, or is that bias towards one or other sex present in many other families? To answer that question I needed to go back to the libraries and the scientific literature.

In fact, the systematic study of the sex ratio in large families began a surprisingly long time ago. Between 1876 and 1885 the German scientist Arthur Geissler examined the birth records of a million families from Saxony with almost five million children among them. Geissler was greatly helped in his research by the rules of registration in Germany at the time, which stipulated that parents had to state the sex of all their existing children on the birth certificate of each new baby. Geissler's was an enormous study by any standards, and all the more impressive as it was accomplished way before computers could help him. But what made it particularly valuable was the huge numbers of large families it included. It would be extremely difficult, if not impossible, to reproduce this study nowadays, at least in Europe, where family sizes have shrunk dramatically over the last hundred years and it is unusual to find a family with more than six children. Far more children were born in nineteenth-century Germany, and far more died in infancy, than nowadays and Geissler was able to find almost two hundred thousand families with six or more children, including an astonishing six thousand families who, like the Lewises, had twelve children.

Geissler himself remarked how often these very large families tended to have far more children of one sex than the other. But was this just the hand of chance or was there an underlying pattern? He was working at a time when, although the laws of chance were well understood, statistical tests of significance had not been properly worked out. While Geissler realized that there were more same-sex sibling groups than there should have been, he was not in a position to know how much importance to attach to these deviations from what chance alone would predict. Geissler published his results in 1889 and for the next seventy years his invaluable records were scrutinized by generations of mathematicians. Geissler was criticized for not taking twins into account (which are more likely to be of the same sex); there was a suspicion that he had unwittingly counted families more than once, and even the implausible suggestion that German parents could not be trusted to give straight answers when filling in forms.

The first scientist to apply statistical methods to this vast set of data was Corrado Gini, who made it the topic of his doctoral thesis at the University of Bologna. Awarded his doctorate in 1905, astonishingly, Gini was still publishing on the topic almost fifty years later when he wrote a magisterial review of the numerous re-evaluations of Geissler's material which had occupied statisticians for the first half of the twentieth century. They had all made various adjustments and added mathematical refinements, but they all supported Geissler's original hunch: that there really were some families which were predestined to have more children of one sex than of the other. The intuition we have that there is something at work here other than pure chance, the flip of the coin, is borne out by the facts. But how does it work? Are there Y-chromosomes around which succeed in biasing the sex of offspring in their favour? If there were, then it might explain how the Sykes chromosome, and others like it, had done so well. Equally, there might be mitochondria – which, remember, do not get passed on by sons – that had managed to bias the sex ratio towards females to enhance their own genetic survival. Are the families, like the Lewises, where one sex is born in preference to the other the extremes in which one sex has managed to establish a complete control? More particularly, is this tendency inherited? If there exist Y-chromosomes, or mitochondria for that matter, that are capable of manipulating the sex of the pregnancy to their own ends, then I would predict the answer to be 'yes'. Unfortunately, Geissler's magnificent set of data is no help in addressing this question, for he only included children from one generation. Getting the answer had to wait for another fifty years.

In the first years of the Second World War the psychiatrist Eliot Slater, working at the Sutton Emergency Hospital on the outskirts of London, was aware of Geissler's work and decided to interview the patients in his care to find out if he too could discover families with large imbalances in the sexes. This was a military hospital and the inmates were mainly soldiers admitted for a range of psychiatric problems. Between 1939 and 1941 Slater and his assistants, a Miss Brown and a Miss Robertshaw, interviewed over a thousand soldiers and asked them the sex of their siblings, children, nephews and nieces. Anticipating the criticism that these patients were not to be trusted to give accurate replies, Dr Slater assures us, in his unfortunately titled scientific paper 'A Demographic Study of a Psychopathic Population', that Misses Brown and Robertshaw were 'clearly aware that doubtful details must be marked as such'. Slater soon found that, like families in nineteenth-century Saxony, the families of British soldiers were also skewed in favour of one sex or the other. Since he also knew the sex of his patients' children and their siblings' children he was able to detect any inherited tendencies. And that is exactly what he did find.

If a soldier had a lot more brothers than sisters he was more likely to have sons than daughters himself. His brothers also shared the same tendency, and had more sons than daughters. These were, in Slater's words, 'male' families. Their Y-chromosomes were doing very well. The reverse was also true, though to a less marked extent. A soldier with a lot more sisters than brothers, from a 'female' family in other words, had more daughters and nieces than sons and nephews. Slater had not only reproduced Geissler's main conclusion, he had shown what others had only suspected – that the predisposition to produce one sex or the other was itself inherited. Slater's natural conclusion was that in any couple the sex of the children is influenced by a combination of the inherited tendencies of each parent to produce boys or girls. These, he suggested, could either work together to exaggerate the bias or cancel each other out so as to equalize the sexes of the children. Under Slater's scheme, a husband and wife who were each from a 'male' family would be more likely to have sons, whereas if both parents were from 'female' families there would be more daughters. A 'male' husband and a 'female' wife (or vice versa) would cancel each other out and have balanced families of sons and daughters.

Slater, and the other researchers who preceded him, had shown that there was some substance behind the widespread intuition. Though they had not realized it, they had glimpsed the battle lines of the combatants. For 'male' families, substitute 'selfish Y-chromosome', and for 'female' families, substitute 'selfish mitochondria', and the opposing elements of the struggle appear from behind the pedigrees and the statistics. What emerges is a world divided into men with Y-chromosomes of various degrees of selfishness – or perhaps 'strength' would be better – and women with 'strong' and 'weak' mitochondria. At the ultimate extreme, the strongest, most superselfish Y-chromosomes would vanquish the influence of all mitochondria in their mates and produce generation after generation of sons. Equally, women with the female equivalent 'supermitochondria' would give rise to generation after generation of girls. Perhaps that is what was happening in the Lewis family from Bournemouth.

In my search for the ultimate superselfish, son-only Y-chromosome, I have come across only one scientific paper which describes such a family. If you are thinking that a man with the ultimate Y-chromosome is bound to be six foot two with rippling biceps and six-pack abs, then prepare yourself for disappointment. This extraordinary pedigree came to light just after the Second World War when the eminent medical geneticist Harry Harris was approached by a man who told him that his family only ever produced boys and asked whether, since he had already had a son and wanted a daughter, anything could be done about it. Was there a chance of having a girl child? The man (we will call him Jack) was twenty-two and, according to Harris's notes, 'thin, quietly introverted and inclined to be solitary; he is untidy, forgetful and has a tendency to daydream'. Jack told Harris that he came from a long line of watch- and instrument-makers and that at least one son in each generation had been in the trade since 1605. All births and deaths in the family since 1690 had been carefully entered into the family Bible, which was still preserved.

From this ancient and unimpeachable source Harris drew out the family tree going back nine generations. All told there were thirty-five children, of whom thirty-three were boys and only two girls. Was this the achievement of a superstrong Y-chromosome on the verge of complete domination? The only two girls in the family for the past three hundred years were Jack's cousin, who had died aged two, and his sister, who was still alive. Without meeting her but by subtle and careful questioning, Harris discovered that Jack's sister was very unusual. According to her brother, she had very hairy arms and legs, so much so that she would never appear in public in a bathing costume. The hair on her head, however, was thin and scanty. She was married but had been told by her gynaecologist that she would never have children. Without the opportunity for a detailed physical examination Harris could not make a definite diagnosis, but he thought it likely that her masculine features were caused by an underlying genetic abnormality of some description. Something very strange was going on in this family. The succession of sons was almost complete and the odds of this happening just by chance were remote indeed – more than a million to one against. Of course, it may have been that chance occurrence, just like the Lewis family may have been. After all, events which have odds of a million to one against do occur – once every million times. The odds of winning the jackpot in the UK national lottery are fourteen million to one against, yet someone wins it almost every week. Nevertheless, while there is a danger of reading too much into this family, there is perhaps a greater danger of dismissing it as a statistical fluke with no underlying genetic interest.

So far, then, we have signs of 'strong' Y-chromosomes which can manipulate the ratio of the sexes in their favour and signs, though not yet proof, of Y-chromosomes that have so perfected the art that the men who carry them have sons but scarcely ever daughters. Precisely how they achieve this power is uncertain. It is at least theoretically possible for a mother to choose, unconsciously of course, to abort the children of one sex, and this is thought to be the likely mechanism by which the sex ratios of some mammals are manipulated. It is much harder for men to do this, simply because they do not carry the unborn child. But a superselfish Y-chromosome can act only through men. How can it possibly work? The clue to a possible mechanism came from research done on, of all things, deep-sea divers. Two separate studies, one on divers from the Royal Swedish Navy in 1977, the other among Australian abalone divers published in 1982, found a huge excess of daughters born to the male divers. Among the Swedes in the study there were twenty sons and forty daughters, while the Australians had among them forty-five sons and eighty-five daughters. These results are way beyond the realms of pure chance and suggest a biological explanation. But what could it be? A solution appeared when it was discovered that working for several hours under high atmospheric pressures, which the divers had to do, lowers the levels of testosterone in the blood. Could that have something to do with it? Another piece of research, this time on subfertile men who had been injected with testosterone, found that they went on to have far more sons than daughters.

Following up this observation, in the 1990s the UK biologist William H. James stuck his neck out and suggested that there was a link between a man's testosterone levels and the sex ratio of his children. He compiled an impressive collection of anecdotal evidence

which implicates testosterone in fathers, and the hormone gonado-
trophin in mothers, in adjusting the sex ratio. For instance, he cited
research which divides professions into 'male' and 'female' on the
basis of the proportions of men and women engaged in each.
Among the 'male' professions are lawyers, doctors, dentists and
scientists, while 'female' occupations include art, literature, music,
psychology and religion. Accordingly, if both parents were drawn
from the same professional category, the sex ratio of their children
would be skewed in its favour, whereas if the man worked in a
'male' profession and the woman in a 'female' one, their influence
would balance out and the sex ratio of their children would
be normal. The only two professions where there was also
information on testosterone levels were physicians, a 'male' profession with high testosterone levels and a high proportion of sons,
and ministers of religion (a 'female' occupation), with lower levels
of testosterone than physicians – and significantly more daughters.
The research James quotes was done over twenty years ago and it
would be very interesting to discover whether the changing com-
position of those professions that were once staunchly 'male' is
having an influence on the sex of the children born to modern-day
practitioners.

Aware as I am of the uncertainties of linking testosterone to sons
as the means by which Y-chromosomes exert their power, there are
some other anecdotes which weave their way into the picture in an
enticing way. For example, it is widely known that the proportion
of sons born goes up after wars. It isn't a massive increase, but it is
real. Immediately after the First World War the sex ratio went from
an average of 103.5 boys to every 100 girls to 106 boys to every
100 girls. The same happened after, and actually during, the
Second World War. This is a real classic for 'good of the species'
enthusiasts, who see it compensating for the number of men killed
during the hostilities – even though the boy children would be at
least twenty years younger than the husbands they were born to
replace. Although the data are beyond reproach, no explanation

was offered by the original researchers for this effect. Indeed, the authors of the report on the US military statistics for the Second World War could not improve on Dr Arbuthnott's conclusion of three hundred years earlier – Divine Providence. William James, however, does have an explanation. People are having sex more often during and after wars. It is certainly true that more marriages happen during wars than at any other time, and also that couples have sex most often during the first months of marriage – or at least they did when these statistics were prepared. It is also the case that children conceived during the first year of marriage are more likely to be boys than girls than those conceived in subsequent years. James puts a hormonal slant on these uncontested facts by suggesting that part of the explanation is that lots of sex puts men's testosterone levels up and that is what elevates the number of sons.

Other than newly-weds, another group of men have lots of sex. They are the men whom others envy – the men with harems. Alas, we have no testosterone measurements for the notorious Moulay the Bloodthirsty, Sultan of Morocco, who lived from 1672 to 1727, but we know he had a lot of children – and a lot of sons. Out of 888 children borne by his hundreds of concubines, 548 were boys and 340 girls. He was by any criteria a wealthy and powerful man, and his Y-chromosome reaped the benefits of his indulgence. Among less extravagantly served yet still powerful men, US presidents have fathered more than their fair share of sons. From the first president George Washington to the forty-third, George W. Bush, American presidents have had ninety sons and only sixty-three daughters. Men have always used wealth and power to attract and collect women – and of course they still do. This is no coincidence, no icing on the cake of success; it is the real purpose behind the accumulation of wealth and power in the first place. Throughout recorded history and throughout the world rich and powerful men have amassed vast harems. In the Middle East the Babylonian king Hammurabi had at his beck and call thousands of 'slave-wives'. In central America the Aztec king Montezuma had four thousand concubines. The Indian emperor Udayama collected sixteen thousand women for his exclusive sexual use. The Egyptian pharaoh Akhenaten assembled a mere three hundred and fifty concubines while in China the emperor Fei-Ti enjoyed sex with a harem of ten thousand women. These collections were not there just for sexual pleasure but were more like vast breeding herds of female humans kept to be inseminated by just one male. They were closely guarded by castrated eunuchs and any stray male who was caught in the act was immediately and cruelly put to death.

While it is perfectly true that all of the emperor's genes will benefit from this abundance of subject wombs, the one to gain most, if Moulay the Bloodthirsty is anything to go by, will be the emperor's Y-chromosome in the body of son after son after son. While daughters will have been born, the rules of patrilinear succession ensured that generation after generation of male descendants could indulge in the same excess, to the delight of the same Y-chromosome. The output of these harems was so enormous that we may still be able to discern the echo in the Y-chromosomes of the present-day population. We can certainly see the persistent evidence of Genghis Khan's success all over central Asia.

There may or may not be Y-chromosomes that do well through an intrinsic ability to swing the sex ratio of their children in favour of boys, perhaps helped by testosterone, but the best manoeuvre for ambitious Y-chromosomes is to associate themselves with rich and powerful men. Once they do so, their hosts' success accelerates the process by increasing the number of boys they have. Everything is going their way.

20

THE SLAUGHTER OF THE INNOCENTS

While there are genetic forces which encourage the birth of more sons or daughters, there are also more deliberate ways to make sure of bringing up children of the right sex. In the Punjabi capital Amritsar, in the far north-western corner of India, a large roadside billboard appeared during the late 1970s. It advertised the services of two doctors who offered to terminate unwanted female pregnancies. It was pitched as a service to women who did not want to give birth to a daughter, only to a son. The pressure to have a male heir was so intense, not least because the exorbitant dowry system meant that daughters were seen as a heavy economic burden, that pregnant women were prepared to undergo an amniocentesis to find out the sex of the foetus and to abort their unborn daughters. The blatant commercialism of this service is distasteful enough, but it is only the tip of the iceberg. The dreadful irony of the Amritsar abortions was that the service attracted the attention of the authorities only because of an error. By mistake they had terminated a boy instead of a girl. The enraged parents complained and this led to a newspaper report which in turn caused a heated debate in the Indian parliament. It then emerged that, far from being an isolated incident, what was happening in

Amritsar was also going on in all India's major cities. The only difference was that elsewhere there was no hard sell. The killing service was being conducted clandestinely in major hospitals all over the country. The furore culminated in the health ministry banning amniocentesis for the purpose of finding out the sex of an unborn child in July 1982.

That made the practice illegal but did not alter the underlying motivation, and nor did it put an end to it. Doctors now used a loophole in the law to recommend amniocentesis to women with the declared aim of diagnosing chromosomal abnormalities like Down's syndrome. Since the sex of the foetus is revealed as part of the process and because abortion on demand is perfectly legal in India, the terminations can be carried out anyway – not because there is anything wrong with the chromosomes at all, but because the foetus is a girl. So the practice continues. An investigation by a women's centre in Bombay found out that of nearly eight thousand requests for amniocentesis over a five-year period in the 1980s, only 5 per cent were genuinely for the diagnosis of genetic defects. The rest were surreptitiously intended to discover the sex of the unborn child. Of the foetuses aborted after the sex had been established by the amniocentesis, 99 per cent were female.

This slaughter of the innocents is not confined to murder in the womb. Rather than hiding the act beneath the twin umbrellas of technology and unseen surgery, others have simply killed girls as soon as they were born. There is no need for a sophisticated genetic test, for delegating the diagnosis to anonymous laboratory technicians. The sex is obvious, the verdict swift and the sentence carried out at once. It has been estimated that in China between 1979 and 1984 a quarter of a million new-born girls were killed – just because they were girls. Recent demographic surveys in both India and China have concluded that there are a staggering forty million fewer females in each country than there should be – the missing millions presumably disposed of by a combination of infanticide, abortion and neglect. The depletion is so severe that in some parts of the countryside there are five times as many young men as girls. Wherever there is legislation, as there is in China, to restrict the number of children a couple may have, it is always the girls that suffer.

In India and China, and in many other parts of the world both now and in the past, girls have been killed either surreptitiously while trapped in the womb or immediately they were born as they struggled for their first breath. We all react to these practices with a sense of revulsion. But why is it happening – what are the root causes? Who, or more particularly what, stands to gain from this conscious manipulation of the ratio between the sexes? These brutal practices and the social logic that underpins them have one very obvious beneficiary, one fundamental element whose purpose is served very well by the elimination of girls. And that, of course, is the Y-chromosome.

Procedures aimed at manipulating the sex of a child before conception have a long and undistinguished history. Chinese and Egyptian manuscripts written more than four thousand years ago discuss the matter of the sex of the unborn child. For instance, if the face of a pregnant woman went a shade of green it was certain she would produce a son. According to Aristotle, aligning the bed on a north–south axis made having a son more likely. Direction or at least alignment during intercourse was also a consistent theme from ancient recommendations. If the man lay on his right side during sex and the woman lay on her right afterwards, a son was more likely to be conceived. This particular recipe comes from the Greek philosopher Anaxagoras in the fifth century BC and began a trend that associated right-handedness in all things with having sons. The thinking, if you can call it that, behind the Anaxagoras family planning strategy was that fertilization was all a question of mixing humours – bodily fluids – and that lying on your right side during intercourse would assure that the humour from the right testicle would prevail. Devotees of the Anaxagoras method were even exhorted to tie off their left testicle if they wanted sons.

Despite the apparently fatal blow to this theory delivered by Aristotle when he pointed out that men with only one testicle could father both sons and daughters, the system lost none of its inherent appeal. Two thousand years later, French noblemen intent on producing a male heir gladly sacrificed their left testicle for the cause – but they made sure never to relinquish the one on the right.

Other techniques, equally unsuccessful, claim to be able to separate X- and Y-containing spermatozoa, the rationale here being that a sperm with a Y-chromosome contains less DNA than those of its rivals in the race to the egg that are burdened with a much larger X-chromosome. The difference in DNA content between the two is 3.5 per cent, and if that were to be translated directly into a weight difference and Y-sperm were 3.5 per cent lighter than X-sperm it would be quite easy to physically separate them. It's certainly true that DNA is dense, but the actual difference in density between the two types of sperm – which is what counts – is only a tiny fraction of 1 per cent. Though the list of techniques, most relying on some sort of centrifugation to spin out the denser X-sperm, is long and sounds very exotic – 'ficoll-sodium metrizoate density gradient centrifugation' or 'ultracentrifugation on a discontinuous sucrose gradient', for example – the basic physics means that, at best, Y-chromosomes are only slightly enriched.

But none of this is really central to the crucial point. No matter whether sex manipulation is technically difficult or extremely straightforward, the fact is that it always favours the Y-chromosome. Whether in the clinics of Amritsar, the nurseries of Beijing or the consulting rooms of London, it is overwhemingly the Y-chromosome which gets chosen to survive. Even where there is no conscious attempt at manipulation, couples in the West more often stop having children after the birth of a son than after that of a daughter. Why do we all expect this, think of it as completely unexceptional? And who, or what, is pulling the strings?

21

THE RISE OF THE TYRANT

Clearly something is going very badly wrong. Virtually all deliberate attempts to manipulate the sex of children favour the male, and Y-chromosomes are having a field day. The Y-chromosome has certainly been the winner, but could it possibly have directly orchestrated its own success? When I looked at my own chromosome, a tiny fragment dried onto a microscope slide, it looked utterly harmless – drifting lonely through the generations, the chromosome that decides sex but has been denied the benefits of sexual recombination. All the other chromosomes are allowed to mix, to exchange genes at every generation. How has this lonely dwarf succeeded in becoming the most influential chromosome of them all? How has this outcast claimed the power to force us, men and women alike, to submit so plainly to its will? How has it managed, in its different hosts, to cast the Viking longships into the violent seas of the north Atlantic, to motivate the Mongol hordes and to kill the unborn girls of Amritsar? This is the ugly face of sexual selection and the genetic conflict which pitches Y-chromosomes against one another and all of them against the essence of the feminine. It has utterly shaped our modern world.

To understand the origin of the rise and rise of the Y-chromosome we must travel back to a time before the world was smothered in ice and our ancestors huddled for warmth around the fire. Then the world was at peace. The great globe of Gaia, self-regulating and maternal, drifted through the heavens around its sun. Our ancestors made little impact on this moving sphere. Its thin coating of atmosphere, the parallel of the protective jelly that surrounds the human egg, maintained the vital gases in correct proportion for life in the seas and on the land, just as it had done for hundreds of millions of years. Three million years ago or thereabouts, our ancestors began to fashion tools from the stones that lay in riverbeds, but that development made no great impression. Ancient humans were very few and far between. By small degrees they expanded from their African homeland to the rest of the dry world, still in small numbers and still with little impact on the equilibrium of the great Gaia. One hundred and fifty thousand years ago, humans of our own species *Homo sapiens*, again from Africa, gradually replaced their cousins *Homo neanderthalensis* and *Homo erectus* in Europe and in Asia. This did not even wake the sleeping goddess of the earth, whose breath wafted gently in the breeze as it always had. To her this was just another animal, among many animals, slowly expanding its range. It was certainly an unusual species, its members capable of communicating with each other in ways she had not come across before, but otherwise fairly unremarkable and nowhere common. The great goddess shut her eyes and drifted back to sleep. A slight change in the earth's orbit cooled her skin and she awoke to see the glaciers oozing from the high mountains and ice spreading from both poles across a frozen sea. The new humans were still there, even in the coldest regions, which she thought impressive but still unremarkable. She shivered a little and fell back into sleep. The Ice Age would pass, she knew, without the need for any intervention. She had seen this happen many times before. She would sleep until it was over and wake again, as she usually did, in another hundred thousand years. To her that was as close as tomorrow is to us.

Twenty thousand years into her slumber she woke with a start, grasping her throat, coughing as bitter gases swirled around her. More volcanoes, she thought before her eyes cleared and she was able to focus. But no sharp heat pricked her skin, the usual signal of a large volcanic eruption. At first the world looked much as it always did. The ice had retreated to the polar caps, the glaciers had moved back up into the high mountains and the sea was blue and largely free of ice. So that Ice Age is over, she thought. The deserts were more or less where they had been before, the green of the forests in much the same places. What could it be that had woken her? Then she looked harder. As she looked down she saw that several patches of grey and brown had appeared around her coastlines and along the rivers that drained her continents. They were completely full of humans, those strange creatures she had quietly admired the previous evening. There were millions of them. The patches were split into tiny squares with layer upon layer of what looked like grey and brown stone pierced with square holes covered in something transparent which reflected the sun. Inside, more humans crowded together sitting on wooden structures or walking to and fro. As the great globe turned and darkness fell on these human hives they were illuminated by tiny specks of orange and white light.

Then new creatures appeared, unlike anything she had ever seen before, coming up from holes in the ground or moving away from where they had been waiting unnoticed. They did not walk or run but moved smoothly forward. Each had two bright specks of white light on their heads and two red specks behind. She could see the humans inside these creatures. Were they trying to catch them? She had seen humans hunt down large beasts before – mammoth, bison, reindeer. But it had not been like this. These humans carried no spears, they were inside the creatures' bodies, but they couldn't stop them running away. They moved so quickly. Then the herds became so dense that they stopped moving. Still the humans stayed inside. Were they themselves the prey of the strange creatures, unlike anything she had ever seen in former times, swallowed whole and waiting to be digested? As it grew yet darker, the strange creatures, still with the humans inside, broke away from the herd and headed off at great speed into the surrounding countryside. They did not disperse in all directions across the plains but along what looked like the dried-up beds of rivers. Great streams of moving lights filigreed across the dark land away from the density of the sick orange glow that lit the skies above the cities.

On the other side of the world, in the first daylight, she saw the same lacy pattern surround other dark stains with the same strange creatures moving on them, now towards, not away from the tight grids of layered warren. Around her coasts, great pipes spouted clouds of choking fumes into her precious atmosphere. These too were surrounded by the same strange creatures she had seen moving along the dry riverbeds. Enormous piles of black rock lay all about, fed by much longer creatures moving along twin ribbons that glinted in the sun. What had once been wide plain or dense forest was now divided into squares and rectangles, each one a different shade of green or earthy brown. Even in her deserts a few green circles had appeared. High above the surface, even above the clouds, vast birds, their wings rigidly outstretched, streaked to and fro at enormous velocities, leaving behind them a trail of their own clouds. The humans were inside them as well. Through tiny holes in the sides of these great birds she could see row upon row of them strapped tightly onto ledges and presumably well on the way to being digested inside the craw of the giant bird. What monstrous chicks in what distant eyrie awaited their next meal?

Gaia's bewilderment with the extremely rapid transformation of her earth was understandable. Great regions of the globe had changed overnight, in her terms, from a landscape of completely natural vegetation to one of cities, fields and roads. Whereas, on her last awakening there had been only a few humans – a million at the very most, scattered across the vast continents of Africa, Eurasia and Australia (though not yet in America), the lands were now teeming with six thousand million of them. The rise of states and cities, of industry and urbanization, all the things we now take for granted because we know no different, all happened in an instant on her timescale. If she fell asleep at ten, by midnight everything had changed.

Accustomed to this modern world, we are unaware of just how fast our circumstances have altered since the end of the last Ice Age about thirteen thousand years ago. It's rather like observing our own ageing in the mirror, or the development of our children: we cannot see the small changes that happen from day to day. Only when we look at an old photograph of ourselves or renew the acquaintance of an old friend we have not seen for years or see a teenager we last met as a young child, only then are we struck by how much we, and they, have changed. Thirteen thousand years ago, our ancestors lived much as they had for the previous two million years. That is not to say there were not changes in our species and its antecedents over that period. There certainly were. But the way of life of our ancestors just thirteen thousand years ago was far more similar to that of our more remote ancestors two million years before than to our own today. They both lived by hunting wild game, by fishing, by gathering wild roots, nuts and fruit. Far less happened in the hundred thousand generations that separate these two groups of hunter–gatherer ancestors than in the five hundred generations that separate us from the end of the last Ice Age. One of the very first types of stone tool, the hand axe, remained unchanged in design or manufacture for three hundred thousand years. My two-year-old laptop is already out of date.

What could have triggered such extreme and rapid change? Aside from those who appeal to the arrival of aliens from another planet as an explanation, most agree that the deciding development was the invention of agriculture. Once we had gained control over our food supply, the rest followed. This seems such an inadequate explanation for all the astonishing complexity of today's high-tech, urbanized world. Being able to grow food rather than having to catch it or dig it up might make life slightly more comfortable but surely it cannot possibly have led directly and inexorably to the paraphernalia of modern life in such a short space of time? Why didn't the equally revolutionary discoveries by our ancestors – of the bow and arrow or how to kindle fire, to name just two – lead to equally vigorous and unstoppable change to modern living? All they did was to make it easier to kill animals at a distance, to cook them and to keep warm. They were the direct, predictable results of human ingenuity. They did not give us the Roman empire, fast cars, caviar, champagne, mobile phones, internet banks, subways, guns, rock bands and mechanized warfare. What was so special about agriculture? How and, more particularly, why did this quite benign, bland, even boring change to being able to grow our own food change us and the world for ever? How did it lead so quickly to our modern world, a world that would have been so alien to our ancestors?

The ultra-rapid transformation to the modern world and its continued headlong acceleration towards the unknown has all the hallmarks of runaway sexual selection. From a hard but stable and sustainable existence, I suggest, our species was suddenly thrown into a whirlwind of sexual selection by the opportunities which agriculture suddenly presented. What were these opportunities? The archaeological sequence of events from the first invention of agriculture to the creation of the recognizable antecedents of our modern world is reasonably well known. I will not argue strongly just now for a Y-chromosome motivation behind the original appearance of agriculture, so let me just describe what we know of its origins. When we say agriculture was invented, it cannot have been in the sense that we generally use that word nowadays. It wasn't like the invention of the steam engine or Mr Dyson's cyclone vacuum cleaner – or the discovery of how to light a fire, for that matter. These were more or less instant advances inspired by flashes of imagination or, more likely in the case of fire, by observation, trial and error. In comparison, agriculture was a system which evolved over a long period.

The earliest of several independent centres for agriculture that we know of lay in the Fertile Crescent encompassing all of modern-day Iraq and parts of Syria, Jordan, Turkey and Iran a little over ten thousand years ago. By then our ancestors had reached all but the most inaccessible parts of the world. They were hunter–gatherers, following the seasonal movement of game animals, collecting the natural harvest of the land and building up an intimate knowledge of the plants and animals, the climate and the landscape. They had crossed the Bering land bridge from Siberia to America and navigated the sea passage to Australia and New Guinea. Only Madagascar off the coast of Africa, the remote islands of Polynesia and, in the northern hemisphere, Iceland and Greenland, remained undiscovered by humans. There never has been a completely satisfactory explanation of why agriculture began in the first place, but the circumstances are known well enough. The sea levels were rising, with water from the melting ice caps and glaciers pouring into the sea as the climate warmed and the Ice Age came to an end. This was not a gentle, incremental process with imperceptible increases in the water level over hundreds of years. Melting of the continental ice caps had created vast inland freshwater seas. One covered half of Canada and the northern United States, its passage to the sea blocked by a plug of ice at the entrance to Hudson Bay. When this barrier finally thawed and gave way, thousands of cubic kilometres of fresh water flooded into the oceans in a single thundering gush. The sea level rose by 8 metres more or less overnight and millions of square kilometres of low-lying coastal plains, once home to bands of our ancestors, were inundated by the sea. By a succession of similar catastrophes, interspersed with gradual rises from unobstructed melting of the polar ice caps, the Persian Gulf was flooded and the people who lived there were forced to retreat before the incoming tide.

They went north along the two great rivers which drained the mountains of Anatolia and northern Iran, the Tigris and the Euphrates, and settled along their banks and in the hills that surrounded the flood plain. A combination of a warmer and drier climate encouraged the spread of wild grasses along the hillsides. Hunters after wild game will no doubt have helped themselves to the ripening seed heads and, equally certainly, someone will have noticed how seeds dropped in soil at their encampment sprouted after rain. It was only a small step – small, but revolutionary – from this chance observation to the deliberate planting of wild grasses. At first the cultivated grass was used just as a complementary food source to add to a diet rich in pistachio nuts and the meat of the migrating Persian gazelle. It made sense to spread the risk in case one food source failed. But whoever it was that deliberately planted the first seed could have had absolutely no idea what he or she had unleashed.

In other parts of the world, in India, China, West Africa and Ethiopia, New Guinea, Central America and the eastern United States, the same thing happened over the next few thousand years. There were different crops in different places, to be sure – rice in China, sorghum in West Africa, taro in New Guinea, maize in Central America and squashes in the eastern United States – but everywhere followed the parallel pattern of cultivation. Agriculture began as a gradual process, at first augmenting the wild food, then in time replacing it. By a similar process, the gradual absorption of wild animals led to the first domestications. It is not hard to picture its first beginnings. A young wild goat, its mother killed by hunters, follows them back to the camp bleating pathetically. Most of the time it would have been added to the menu, but it isn't at all hard to imagine a child, accompanying his father on the hunt, wanting to keep the young animal as a pet. After all, children still do the same today with young birds or animals found wounded by the roadside. There is no need to give our ancestors the credit for predicting the outcome of such a small act of kindness. The goat would be perfectly happy tethered to a tree close to the encampment and feeding on whatever it could reach. It would not have been so happy when it grew up, lost its appeal and was eaten for supper. It is a very small step from there to deliberately taking young animals from the wild and breeding them in captivity. Neither the gradual domestication of wild grasses nor the progressive taming of wild animals seems such a great shift in the individual lives of our ancestors. Hardly sufficient, one would have imagined, to be the catalyst responsible for all of the wonders and the terrors of our modern world. And the archaeological record shows that the agricultural way of life spread only very slowly. From its beginnings in the Middle East it took over four thousand years to reach our ancestors in Europe and change their way of life. On the coasts of Denmark, for example, people went on living well for more than a thousand years on fish and shellfish which were everywhere abundant, while their neighbours 50 miles inland cultivated fields of barley.

While I don't seek to underestimate the direct benefits to our ancestors of having a way of supplanting their wild food with home-grown crops and animals, there was another side to the new skill – an effect which reverberates right down to the present day. The invention and adoption of agriculture was accompanied by new concepts with a far greater lasting consequence, concepts which were unknown before the first seed was planted or the first animal tethered to a tree. These concepts were property, wealth and power. They were entirely new and played straight into the hands of our old friend – the Y-chromosome – as a new and irresistible instrument for sexual selection. Now, at long last, there was an opportunity for Y-chromosomes that could get hold of these valuable assets to increase almost without limit; an opportunity to pursue their natural instinct for endless replication that had until then been contained. It was, in my view, men and through them the Y-chromosome that seized on this trio of property, wealth and power and pushed them to their present absolute prominence. It may even be that this seductive combination, coupled to the unstoppable force of sexual selection, was not the passive and innocent by-product of agriculture and husbandry but the driving force behind its spread around the world. With property, wealth and power to play with, the Y-chromosome suddenly found a way not only of beating its rivals, other Y-chromosomes, but of crushing its age-old enemy – the mitochondria, guardians of the feminine. Innocent agriculture was the key that unlocked the chains that had restrained the raging beast of Adam's Curse and let it loose upon the world.

Before ten thousand years ago there was no wealth, no personal property to speak of and certainly no ownership of land. Bands of our ancestors moved through the landscape following the annual migrations of their animal quarry, making sure they were in the right place at the right time. Their shelters were for the most part temporary and seasonal. Sometimes they camped at a convenient ambushing point, perhaps a river crossing, where they knew of old that the herds of bison or reindeer would cross in the spring and again in the autumn. In summer they might be higher up in the hills to collect the eggs of birds or catch fish in mountain streams. Winter would find them on the coast, digging up shellfish from the sand or trapping shrimps in tidal pools. Through these wanderings, our ancestors were never separated from the land. They knew every plant, whether it was edible or poisonous and whether it had special properties as a painkiller, hallucinogen or aphrodisiac. They knew the animals, the birds and the fish. They knew which ones to avoid, how to stalk others, and when and where to trap their food. All these things our ancestors knew. All these things we, their descendants, have forgotten.

Our ancestors, whose genes we carry, were part of the land. They lived so recently that their genes sit unchanged within us still, and call us back to the wild from time to time, to the hills and to the sea. How easily we learn how to move quietly along a riverbank and cast a fishing fly to a trout, instinctively taking advantage of the natural cover. How suddenly, walking through the woods, we stop as the scent of a fox drifts across the path. How we hurry home as darkness falls, away from the dangers of the night when leopards and other predators roamed. How we light a fire for reassurance as much as for warmth and how, as it crackles into life, it becomes the centre of the house and we feel much safer. These are the echoes of our ancestors carried to us by the genes they gave us. Genes that favoured caution, bravery, patience and hunting skills. Do not be a bit surprised to feel these atavistic tugs, now almost lost in our subconscious, from time to time. It is through these instincts that we sense the threads which connect us to our ancestors, whose world we know and yet we do not know.

The greatest casualty of all is the fission of the sexes. The blind rage of the male, released from its chains, has slowly and deliberately enslaved the female. But how did agriculture catalyse this utter transformation of human behaviour and erode the balance between the sexes which had for so long sustained our ancestors? The early farming settlements of the Middle East give us the archaeological clues – places like Jericho in the Jordan valley, which was continually occupied for over eight thousand years until relocated by Herod just over two thousand years ago; or Abu Hureyra, an even earlier settlement in what is now Syria; or Catal Hüyük, the later farming village on the plains of Anatolia in modern Turkey, dating from between 8,200 and 7,500 years ago. In each of these locations the evidence is there of the rise of ownership, the suppression of women, and our gradual separation from wild nature. In Abu Hureyra, for instance, the skeletons of women betray the evidence of their domestic enslavement. They show the unmistakable signs of osteoarthritis, damaged vertebrae and curvature of the thigh bone coupled with bony outgrowths of the kneecaps, all injuries consistent with a life tied to kneeling at the grindstone. Our ancestors rapidly lost their intimate knowledge of the multitude of wild plants built up from millennia of harvesting what grew in the wild. Plant remains found in Jericho show that the inhabitants quickly came to depend on only a dozen or so cultivated plants, foremost among which were wheat and barley.

By selecting and planting the grains that were both the plumpest and the easiest to harvest the following year, our ancestors began unknowingly to replace natural with artificial selection. By this means, the characteristics which most suited the farmer were retained, and soon the cultivated wheat and barley no longer resembled their wild antecedents. Exactly the same process of artificial selection rapidly changed the ferocious aurochs, wild cattle standing 2 metres tall at the shoulder, into pliant and manageable domestic cows. Farmers erected fences to contain their animals. The animals *belonged* to someone. Farmers felt they *owned* the land they had cultivated and the seed they had stored. These were concepts of property unnecessary and unknown to their hunter–gatherer ancestors. Though they too stored food, made tools and weapons for their own use, and traded raw materials and finished goods with neighbouring bands, this was on a small scale just sufficient for survival. Ownership of animals and land were completely foreign concepts for hunter–gatherers, as can be seen by the ease with which native Australians and other aboriginal peoples were swindled out of title to their ancestral lands by Europeans well schooled in the economics of wealth and property.

The settled life brought other major changes that were to strain the bonds between men and women. The constant movement across the landscape of the hunter–gatherers as they moved from one seasonal camp to another imposed strict limits on the spacing between children. It was quite impossible to contemplate having a second baby while the first was unable to walk well enough to keep up with the rest of the band. Children were not weaned for three or four years because their mothers could not ovulate while they were breast-feeding. They thus avoided a second pregnancy until they could be sure the first child was fully mobile. Agriculture changed all that. Because our ancestors were no longer continually on the move, there was no longer an absolute requirement for such long gaps between pregnancies. That at first might seem like a positive benefit, but it proved to be the very worst thing that could have happened to women. Instead of just enjoying the rest and relaxation which a sedentary life had to offer, women were forced to reduce the spacing between births from four or five years to one or two years. Forced by the relentless ambition of the Y-chromosome to reproduce itself, women were reduced to a state of serial pregnancy, increasingly enslaved by dependence on men.

This suited the Y-chromosome as the sexual landscape turned in its favour. The irresistible opportunity arose to build a harem, a herd of women just as dependent on its owner as his sheep or cattle. Women themselves became domesticated and imprisoned. The temptation to polygamy was overpowering and examples were all around. Men, driven on by the lash of their Y-chromosomes, could copy their cattle and become the stud bulls of their own herd. But the damage didn't stop there. The enslavement of women through serial pregnancy required much earlier weaning than before. No longer required to be able to walk and run before being released from the breast, the young child needed to be weaned. Some archaeologists believe this was accomplished by the invention of fired pottery which allowed cereal grains to be boiled into a pasty gruel which could be fed to unweaned infants. Once her child was weaned, a woman could become pregnant again soon afterwards. The bull/man would have no difficulty at all with that part. But the children, ripped from the security and unconditional love that breast-feeding embodies, were left feeling bewildered and abandoned. Far from gaining a sense of independence, they were bereft, deprived of the strong sense of their own value and autonomy which builds during this intimate and prolonged contact. Some believe that children even now never really recover from this shock. They struggle to regain trust in a world that has for some reason unknowable to them abruptly changed for the worse. The trauma of early weaning has even been adduced in modern theories of depression. The feeling of powerlessness implanted by the sudden withdrawal of love and nurture at the mother's breast, when even the cries of despair go unanswered – as they must for early weaning to succeed – leaves a long shadow in the psyche of the very young that can darken their whole lives. Agriculture also increased the demand for children. They could be put to work in the fields where there were many unskilled jobs which needed doing. Previously they had to pass a long childhood before they could be taken to the hunt. But now they could be put to good use more quickly, increasing yet again the wealth of the family – which meant, of course, the wealth of the man. In time, others were likewise enslaved to work the fields and maintain the herds. The agricultural way of life created inequality: inequality between men and women and inequality between the wealthy and the poor; between those with land and animals and those without it. Social stratification made its first appearance. Men without land or animals sank to the bottom, forced to work for their wealthier neighbours. And these neighbours also, of course, collected up all the women.

Less tangible, far more debatable but nonetheless fascinating are the echoes of the past contained within our own mythologies. They are not, of course, literal truths and they emerge like ghosts from the past, stories which connect us to the world of our ancestors. They were not written down until well after agriculture had become established but many hint at a change, a shift at around that time from a matriarchal theology to one dominated by men. The earliest art, the 'Venus' figurines of the Upper Palaeolithic, some 25,000 years ago, are small statues moulded from clay or carved from stone, and are all of women. Do they represent, with their often exaggerated breasts, the image of a Great Goddess? Stories of such a Goddess percolate down into recorded history as half-remembered fragments of much earlier times. The admittedly idiosyncratic interpretations of several creation myths by Robert Graves conjure up an image of a time when there were no gods or priests but only a universal goddess, supported by her priestesses. According to Graves, woman was the dominant sex and man her frightened victim. The mystery of childbirth was a secret among women and men were ignorant of the part they played in conception, fertilization being attributed to the wind or the swallowing of an insect. The sudden reversal, the new world order, the abrupt change to patriarchal theologies was marked first by the Babylonians, then by the Greeks and after them by Judaism, Christianity and Islam, whose Creator figures were always male.

This switch began, according to Graves, when men realized that it was they, rather than the wind, that could claim the credit for initiating the birth of a child. Men were no longer recruited merely to entertain the Goddess, and when the tribal Nymph selected a young man as her lover he also became a symbol of fertility – although he was generally sacrificed at the end of the year. By various ruses these consorts put off their execution and ruled with the Queen. By the time the Hellenic myths came to be known, men were firmly in the driving seat. The demise of the goddess was symbolized by the Babylonian god Marduk, who killed a dove, symbol of the creator/goddess Iahu, at the Spring Festival, and by the Greek hero Perseus, who beheads the goddess Medusa. Make of these uncertain signals what you will (some feminists have objected to this interpretation as an excuse for modern patriarchism, as if it was the misbehaviour of women that caused them to be overthrown), but the deep-seated feeling that women are truly goddesses and men's principal purpose is to worship them still hovers uneasily in the collective subconscious.

The inequalities between the sexes did not escape the notice of the Great Assembly of genes, the nuclear chromosomes. Indifferent to which sex transports them to the next generation, they began to savour the prospect of being carried along by wealthy men with their new opportunities for polygamy. The train of sexual selection was gathering speed, the boilers stoked by the energy and ambition of the Y-chromosome, the Great Assembly waving it off from the station. Just as power and wealth converged on fewer and fewer men, so their wealth became more and more necessary to the survival of the women, now utterly dependent and suppressed.

Chiefs emerged, villages coalesced into small states, tribal groups grew together. Wealth and power, the only things that mattered now, replaced the virtues of the hunter which had earlier guided a woman's choice of mate. Women now chose, where any choice remained to them, on the basis of wealth and property. The runaway train of sexual selection was by now thundering along the tracks. A man with wealth could expect to have more wives or, failing that, more women to inseminate. Driven on and on by the crazed ambition of the Y-chromosome to multiply without limit, wars began to enable men to annex adjacent lands and enslave their women. Nothing must stand in the way of the Y-chromosome. Wars, slavery, empires – all ultimately coalesce on that one mad pursuit.

Our recent history is a catalogue of greed and domination, a conspiracy to which we all subscribe, men and women alike. We are now all so thoroughly marinated in the juices of possession, money and property that we are blinded to the ultimate destination of the runaway trains. Somerled, our hero of an earlier chapter, is typical of the successful tyranny of the Y-chromosome. He is valiant, brave, defender of his fathers' lands. We are all instructed by his example and inclined to admire him for his manliness, his heroism. The trail of destruction and slaughter his Y-chromosome leaves behind we all glorify in verse and myth. He killed the first man he saw and ripped out his heart. What a man! But even the havoc and carnage wrought by this strictly local hero would not be enough to trouble Gaia. She would hardly notice.

The limits to sexual selection imposed on the animal examples we looked at in an earlier chapter are reached only when the adornment is so disadvantageous that it becomes a burden – the huge male elephant seal that is so heavy it cannot get onto the breeding beach, or the peacock whose tail is so splendid and so large that it cannot fly away from predators. But there is no natural limit to human sexual selection based on wealth and power. There is no negative feedback control. Wealthy and powerful men are not disadvantaged. They generally get richer still. The mad scramble, fuelled by the most basic of unseen genetic impulses, seriously endangers the survival of the species – and the planet. In ten thousand years we have changed from an intelligent and resourceful animal, quite rare but with remarkable skills and a natural part of Gaia's world, into a teeming species very rapidly destroying her beautiful planet.

In its latest report, looking forward to 2003, the Worldwatch Institute, an organization based in Washington DC which monitors the deteriorating natural and human environment, predicts a future of continuing misery and biological impoverishment. Already 1.2 billion people, one-fifth of the world's population, live in absolute poverty, defined as surviving on less than a dollar a day. Global warming is accelerating and the atmospheric concentration of carbon dioxide, the product of burning fossil fuels, has reached levels not seen for twenty million years. Melting ice is causing sea levels to rise – by an anticipated 27 centimetres over the next hundred years. Thirty per cent of the world's surviving forests are fragmented and being cut down at the rate of 50,000 square miles a year. A quarter of the world's mammals and one-eighth of its birds are in danger of extinction – fifty times the natural rate. Industrial pollution has reached all-time highs and toxic chemicals are being released in ever-increasing quantities with only the vaguest idea of what damage they do both to humans and to natural systems. We all know this, yet we ignore it. We know we should stop producing so much carbon dioxide. We know thousands of nuclear warheads are stockpiled around the globe. We know we are pumping toxic chemicals into the oceans and filling the skies with poisonous gases. We know we should stop. But we cannot. The runaway train of sexual selection is gathering speed and, with the blind Y-chromosome in the driving seat, completely unaware of these extreme global dangers, it races on out of control. Unless something happens it will leave our beautiful planet not just dying but dead: another lifeless rock spinning round the sun.

22

THE SPERM OF TARA

I have painted a very black picture of a world driven by the coupling of sexual selection working through its new playthings – wealth, power and greed – hand in hand with the Y-chromosome to deliver the present nightmare of patriarchal dominance, misery, poverty and destruction. Now you know why I called my book *Adam's Curse*. How is this different from just blaming men for everything – a common enough complaint? The difference is that sexual selection involves *both* sexes. Only if wealth, power and status in a man succeed in 'persuading' more females to mate with him than with his rivals will it work, let alone build up its present momentum. I say 'persuading' knowing full well that Genghis Khan used an army to persuade and female choice didn't come into it in any real sense. And, of course, Y-chromosomes, intent on ensuring their own survival through sons, cannot be *blamed* for anything, any more than mDNA could be *blamed* if it organized its own preferential survival through daughters. In fact, mDNA is in a much stronger position to do that, and there is evidence building up that mDNA, the essence of the feminine, is quite capable of fighting back. For the first signs of this counter-trend we must leave the world of humans for a moment and return to the world of insects.

242

Remembering that cytoplasmic genes and mDNA have no interest in producing sons and that their own long-term future lies only in producing future generations of daughters, can they do anything about it? If mitochondrial DNA and other cytoplasmic genes are forced to endure sex, which does them no good at all because they don't get to enjoy the benefits of recombination, can they fight back by killing or disabling males? Yes, they certainly can. The first example of the deliberate slaughter of males was noticed in 1975 by Sir Cyril Clarke. He was an extraordinary man by any measure. His day job, as it were, was as a professor of medicine, and he was one of the very few doctors who took any notice of genetics before the molecular revolution of the past twenty years. His great medical achievement was to find a cure for haemolytic disease of the new-born, an often fatal disease caused by an incompatibility between the Rhesus blood groups of a mother and her unborn child. He eventually became the President of the Royal College of Physicians and died, aged ninety-three, in 2001. But, as well as his medical accomplishments, he also led a parallel life as an entomologist and was renowned as a skilled breeder of butterflies and moths. His entry in *Who's Who* listed breeding swallowtail butterflies among his hobbies.

In this other life, Cyril Clarke's research centred on the genetics of mimicry. Poisonous insects are frequently brightly coloured to warn their avian predators that a nasty surprise awaits should they decide to attack. Cunningly, other butterflies have evolved to mimic the poisonous species but without going to the trouble of developing their own toxins. This cloak of deceit does indeed fool the birds, but only up to a point. If the garish colours become too common birds quickly learn that the vivid pattern is only a pretence, not backed up by the punch of real poison, and become prepared to take the risk. So the mimics have developed two alternative outfits – one garish and pseudo-toxic, the other camouflaged. The proportions of the two genetic forms, which can look entirely different, are beautifully balanced in the wild. There are enough individuals with the vivid outfits to remind the birds who they are, but not too many to spoil the subterfuge. One of the butterfly species which interested Clarke was *Hypolimnas bolina* from the forests of Queensland, Australia. The males, with stunning blue and white ocelli, or eye-shaped spots, on a deep black background, are all identical and only the females exist in different, mimicking forms. When he began to breed this butterfly, Cyril Clarke noticed that females captured at a certain location produced only female offspring. Half the eggs failed to hatch. When he examined them under the microscope he noticed, from details of their internal structure, that the unhatched eggs were all male. By breeding experiments between butterflies from this male-killing strain and others from normal strains, Clarke showed that, whatever was killing the males, it was being inherited through the females – just like the cytoplasm. The male-killing females needed to mate with males before they could lay fertile eggs, but something in the cytoplasm was silently murdering all their sons.

Laurence Hurst, the champion of the cytoplasm, found another example in the two-spot ladybird *Adalia bipunctata*. These are the familiar little red-and-black insects commonly found in gardens whose larvae feed voraciously on another of the characters we have met already – the aphid. By breeding experiments similar to Clarke's, Hurst discovered that the *Adalia* females also managed to kill male eggs by something passed on through the cytoplasm. This time, though, there was a bonus for the female young. They could feast on their slain siblings, silently killed before they had even hatched. The butterfly and the ladybird were the first to lift the veil on the revenge of the cytoplasm. This was not an open blood-and-guts campaign against males but a silent elimination perpetrated by the favoured means of the female – poison.

For the most chilling example of cytoplasmic revenge and assisted murder we turn again to William Hamilton and his research on what became his all-time favourite insect – the parasitic wasp. These often tiny wasps lay their eggs in the larvae or pupae of other insects. The eggs hatch and the larvae devour their hosts from the inside. They quite literally eat them alive. One of these wasps, the minuscule *Trichogramma*, only 1 millimetre long, lays its eggs inside the eggs of butterflies and moths. When the larvae hatch they devour the contents of the egg and then pupate before hatching as adult wasps. But there were never any males. *Trichogramma* appeared to have given up sex altogether, producing brood after brood of females with not a male in sight. On the face of it, this was simply another example of a species abandoning sex – just like the dandelion. But, in a remarkable experiment, Hamilton showed that the wasps hadn't given up sex permanently at all. If he fed the wasp larvae on honey containing a strong dose of the antibiotic tetracycline, they reverted to a sexual life cycle. Both males and females were born from these larvae and, after a few generations of larvae had been reared on the antibiotic-laced honey, they were indistinguishable from a regular sexual species. When the antibiotic was withdrawn, the wasps carried on as if they had forgotten their generations of celibacy and had males, females and sex in the normal way. What strange magic was at work here? It turned out that bacteria, carried in the cytoplasm, were manipulating the sex of the offspring not by killing males but, incredibly, by turning male embryos into females. After a few generations of dosing with tetracycline, the bacteria were eliminated, the sex change reversed and a fully sexual way of life resumed. Being confined to the cytoplasm, the bacteria – just like the mitochondria – had no interest in producing male offspring. Whether the bacteria were the primary manipulators in this case or whether they were merely the assassins hired by the mitochondria to eliminate males, indeed to eliminate sex altogether, has not yet been determined. But both bacteria and mitochondria have precisely the same interest in the outcome. Kill males and succeed. In some species, as we have seen, this strategy has been taken through to its ultimate conclusion. Males have been eliminated altogether and the females just carry on cloning.

These are subtle strategies, not the blood-and-thunder shoot-outs that we expect from the Y-chromosome. Might we find anything like them in our own species, and where would we look? Just as Cyril Clarke noticed among those butterflies that produced only females, that could be the place to start in humans – families with a record of producing only daughters. We have met the Lewises already, with their twenty-three girls and just four boys in the maternal side of the family. Tracy Lewis's mDNA is doing very well, but if it has managed to outwit the Y-chromosome and have mainly girls, still it has not quite perfected the art. Ironically, the Y-chromosome itself makes it difficult to find families who might have discovered the secret of nudging males aside. Without the assistance of a surname to draw attention to their success, as it did for Sykes, Dyson and Macdonald, candidates for top mDNA are much more difficult to discover. No simple inspection of school records is enough to reveal a tendency to produce daughters for the simple reason that women have generally changed their surnames at every generation. This practice, as any genealogist will tell you, is the greatest single obstacle to reconstructing maternal genealogies from the records of births, marriages and deaths. There is no simple way of teasing out those maternal lines, those lines of mothers who manage to have more daughters and fewer sons, from the records. Only dramatic examples, like the Lewis family, exist to hint at the possibility and they come to light only for exceptional reasons. If Tracy Lewis had had only a small family of two or three girls, she might still have possessed the same ability to nudge males aside but no-one would have noticed. Indeed, the most spectacular example of a superselfish mDNA came to light quite by chance.

In 1947 a woman was admitted to hospital in the French city of Nancy, provincial capital of the *département* of Meurthe-et-Moselle in north-eastern France, 100 kilometres to the west of Alsace. She was there so that she could be kept under observation during the last few weeks of her pregnancy because she had lost her first baby to a late miscarriage three years previously. The pregnancy continued without any difficulties and her child was duly delivered perfectly normally. When the doctor delivering the baby announced it was a girl, the woman seemed to be completely unsurprised by this news. 'Of course it's a girl,' she replied, 'my family produces only girls.' This must be a relatively common occurrence, for there are bound to be families with histories of producing daughters just by chance. But what makes this case at first unusual and then utterly remarkable is that the physician actually followed up this casual remark. What he found was not just a tendency in the family to produce girls – it was an absolute refusal to have any sons at all. Tracing the woman's ancestry back, he discovered that she had an astonishing total of seventy-eight maternal relatives over nine generations. Seventy-eight daughters and not a single son! The odds against that happening as the result of the coin-tossing random process of deciding on a child's sex we looked at before are more than a hundred million to one. Of course, there is bound to be someone who will say that in a world of sixty thousand million people, things happen with odds against of a hundred million to one all the time. But I'm impressed.

Though they do not share the same surname, as they would if surnames were maternally inherited, these women are connected by their mDNA. This tiny circle of DNA would trace the family from Nancy back through nine generations of maternal relatives as easily as Somerled's Y-chromosome connects the chiefs of Clan Donald. This, it appeared, was a mitochondrial lineage that had found the secret of eliminating the chore of having sons. But how? Had these women found a way of refusing to have their eggs fertilized by sperm containing Y-chromosomes or of neutralizing the sex-determining gene altogether? Or had the mDNA subverted the implantation mechanism so as to reject all male embryos or, most gruesome of all, arranged to abort all male foetuses? Sadly, the trail has gone cold; the family is lost and these questions have not been answered. But what, I wonder, was the sex of that woman's unborn child, aborted just a few weeks before being born? Was he a boy unconsciously murdered in the womb? We have seen this happen in insects and dispassionately debate the reasons why. But how we recoil when the ruthless hand of evolution shows the same phenomenon in ourselves.

How had this maternal lineage, followed if not directed by mDNA, managed so absolutely to prevent the birth of children in which it had no possible future interest? Only daughters pass on mitochondrial DNA. Sons do not and so are merely a tiresome burden. How had the lady from Nancy arranged to eliminate all Y-chromosomes? Putting aside the possibility that she was, like the summer aphid, reproducing without sex, there is no doubt that she and all of her maternal relatives were inseminated by sperm of which half contained Y-chromosomes. But none got through. And if the lady from Nancy and her relatives could succeed so spectacularly in becoming the nemesis of the Y-chromosome, how many more women might there be who are not quite so skilled, not quite so practised at killing their sons, but still manage it to a lesser extent? How many wombs invite only to destroy? The nine months a Y-chromosome spends inside the female body, far from being the safe and protected sojourn we all imagine, might be the most dangerous time of its life. There is so little hard evidence to support this idea that I hesitate even to raise it. But if there are Y-chromosomes that have managed somehow to encourage the production of sons, then there may well be a counterbalancing influence which prevents the over-production of sons by encouraging daughters. I do not have any direct evidence for male-killing strains of human mitochondrial DNA, particular mitochondrial sequences which pose a special threat to the Y-chromosomes that look to it for nurture during those crucial nine months. But there is one aspect of my own work on mitochondrial DNA that has always puzzled me and my colleagues and has never been satisfactorily explained.

In Europe, the seven clans, the seven clusters of maternal descendants from the seven ancestral women, are found in every part of the continent. But one predominates in every single country. At least 40 per cent of native Europeans are descended from the clan of Helena, three times as many as the next most frequent clan, Ursula. When I am asked about this at lectures, the question usually comes in the form: How do I know there hasn't been any selection? That is to say, how do I know that the geographical distribution and whether a clan is frequent or rare might not be due at least in part to selection, to some mitochondrial DNA having an advantage over others? It's in some ways a fair question; it is asked with monotonous regularity and always in the expectation that I won't have an answer. And I don't. Scientists are accustomed to thinking about a selective advantage as conferring some sort of material change on the individual who carries it – making it bigger, fitter, more resistant to disease and so on. I personally can't see how mitochondrial DNA is going to make much difference in those leagues – though it is practically impossible to know exactly *how* a selective advantage actually works in practice. But suppose that, rather than making the carrier better able to survive and reproduce in a conventionally comprehensible way, women in the clan of Helena were able to direct the sex of children towards the female rather than the male. Other things being equal, that could be a very substantial advantage for her mDNA. A mitochondrial DNA like that would spread quickly, just like a Y-chromosome with reciprocal properties (producing more sons than daughters). Could that be the hidden advantage to the clan of Helena that had pro-pelled it to compose almost half of all European mitochondrial DNA? Is that the reason why Helena's daughters became so abundant? Not a greater efficiency in metabolism, which would be a straightforward physiological explanation, but a greater efficiency in avoiding sons?

The only fragment of direct evidence for anti-male behaviour by mitochondria that I have been able to track down comes from a paper published in the *American Journal of Human Genetics* in September 2000 that reported research carried out by a team of doctors from Zaragoza in north-eastern Spain. They had been working on infertility in men, a not uncommon problem in many parts of the world, as we shall see. Between 10 and 15 per cent of couples are infertile to a greater or lesser degree, and in roughly half of these couples the infertility is traceable to the man. There are very many reasons why males might be infertile – they may have an extra chromosome or undescended testes, or have been exposed to poisons or radiation. But in more than half of men who seek treatment for their infertility, the problem lies in their sperm. Either there aren't enough of them or they can't swim properly. Among the non-swimmers, the defect in some has been traced to a deletion of segments of the Y-chromosome that has removed genes. We met a similar situation in chapter 5, when we saw how other deletions helped scientists track down the sex-determining gene on the Y-chromosome, and we will revisit this phenomenon later.

But it was not the Y-chromosome that the Spanish researchers had their eye on. It was the mitochondrial DNA of these infertile men. Although mitochondrial DNA, as we know, is passed on only down the female line, and we all get ours from our mothers, sperm do contain a few mitochondria which they need to provide the energy for propulsion. Connecting the head to the tail is the midpiece, which is where the mitochondria, about a hundred of them, are located. Usually only the sperm head, containing the nucleus, enters the egg, but even if a few mitochondria from the midpiece do get through, they are soon identified and destroyed. The mitochondria are there in sperm to provide energy and they contain the necessary enzyme catalysts to do so. If these are poisoned, the sperm stop swimming, proving that mitochondria are vital for sperm propulsion. But could mutations in this essential mitochondrial machinery also be a cause of sperm running out of puff? That is the question the Spanish team wanted to answer. Rather than confining their research to infertile males, they persuaded almost six hundred volunteers from Zaragoza and Madrid to donate a semen sample. Within two hours of the donation, the sperm were thrashing away under the microscope and given a score from A to D depending on their vigour. A-list sperm moved quickly across the slide, their flagellae beating rapidly. B-list sperm still moved, but were more sluggish. C thrashed but didn't move, and D just sat there neither thrashing nor moving. Men with less than half their sperm in the top two categories A and B were classified as, wait for it, *asthenospermic* – literally, weak-spermed.

The inelegant and pedantic next step would have been to try to find DNA mutations in the men whose sperm were tired out. Instead, the Spanish researchers made an inspired move. They checked out from which of the Seven Daughters of Eve the men were descended. That seems on the face of it a very peculiar thing to do, but the reasoning was absolutely ingenious. While, as we saw, any Y-chromosome mutations that slow down sperm will be quickly eliminated by natural selection, the same logic does not apply to mitochondrial mutations with similarly adverse effects on sperm. Unlike Y-chromosomes and the rest of our genes, mito-chondrial DNA does not depend on sperm to get to the next generation. It is passed only through the female line. So a mito-chondrial mutation can make sperm as sick as it likes without in the least affecting its own survival. These mutations, the Spanish researchers reasoned, could and would persist through generation after generation of women. There was no reason for them to be eliminated by natural selection if all they did was to cripple sperm. It might not help the species, but what do they care? If these sperm-slowing mutations had been inherited for generations, men who had them in their mitochondrial DNA might be related to one another through their mothers. What better way to test that idea than to take advantage of the different clusters of mitochondrial DNA that had already been identified in Europe? And that is precisely what they did. By checking a few key sequences that we and others had published in scientific articles, the Spanish team were able to divide their volunteers into the mitochondrial descendants of Ursula, Xenia, Helena, Velda, Tara, Katrine and

Jasmine, the names I gave to the seven European clan mothers. And there they found a connection. Men from the clan of Tara had sperm which were significantly more sluggish than the sperm of volunteers from the other six clans. Spurred on by this remarkable result, the team then challenged the sperm to swim in a straight line up a thin glass tube. After half an hour they stopped the clock and measured how far the sperm had got. Sure enough, the Tara-fuelled sperm came in last, averaging just over 7 millimetres an hour. They lagged almost a whole millimetre behind Xenia, Ursula, Katrine, Jasmine and Velda – but, storming out in front, at an impressive 11 millimetres per hour, was Helena.

This is a fascinating piece of research in many ways, and leads to further intriguing questions. For instance, are the speed and endurance shown by the Helena-fuelled sperm reflections of all-round mitochondrial energy efficiency in the body cells that would help explain the high frequency of the clan in Europe? Or are they peculiar to sperm metabolism alone? And what about the poor Tarans, whose sperm limped in last? Are they equally handicapped in other departments where metabolic efficiency is important? That can't be true of the Taran female descendants, otherwise the clan would never have survived. But Taran men, completely irrelevant to the future prospects of the mitochondria they carry, could be metabolically compromised without in the least affecting the success of the clan. Speaking as a Taran myself, I do hope not – though I do find it devilishly hard to get up in the mornings.

This intriguing experiment of the Zaragoza doctors, with its far-reaching conclusions, is important not just in the context of male infertility but as a demonstration of the disregard mitochondrial DNA has for the breeding success of its male carriers. It has not shown that mitochondrial mutations are directly capable of producing daughters rather than sons. But it does show emphatically that mitochondria are quite able to influence the fertility of sons and reduce the chances of their passing on their Y-chromosomes to the next generation. This counts as a qualified victory in the battle of the sexes. The patient mitochondria wait for a generation before uncapping the poison phial. The poison saps the vigour of the sperm in the most direct of ways, by cutting off the energy supply. This form of male infertility is biochemical, almost mechanical and brutally straightforward.

Might other patterns of male infertility, less obviously clear-cut, be similarly laid at the door of mDNA? Though it is not generally thought of in the same way, there is one other form of male infertility that sprang to mind: that of the male homosexual. Though no gay man is likely to be referred to an infertility clinic, from a purely genetic point of view theirs is a self-imposed infertility. Have gay men, like Taran sperm, been kissed by the same poisoned lips? Surely not? But I was sufficiently intrigued and impressed with the guile of Tara's mitochondria in disabling her sons to think the question worth pursuing further.

23

THE GAY GENE REVISITED

As a geneticist I have been curious about homosexuality for a long time. The curiosity is that, if there is a genetic basis to homosexuality, then there must, by definition, be genes involved. The question is, how do the genes get passed on from one generation to the next? After all, at the simplest possible level, sex between partners of the same sex, while it may be fun, cannot result in children. Sperm might be delivered but it never sees an egg. I am well aware that gay people do have their own children, either by surrogacy or from previous straight relationships. But it is common sense that, taken over all, gay men – and my curiosity is largely focused on them – cannot have as many children as straight ones.

I have worked on inherited diseases for a good part of my scientific career and there is no denying that homosexuality has some of the genetic characteristics that you might find in a serious inherited disease. As soon as I write that sentence I can hear the loud objections ringing in my ears and imagine myself on a television chat show, in front of an audience, accused of saying that homosexuality is abnormal, a genetic disease, and spending the rest of the show on the defensive, denying that I ever said anything of the kind. But I can't help being curious – in fact, I *ought* to be curious – about how a characteristic like homosexuality could possibly be inherited, if indeed it is. The point of comparison with serious inherited diseases is that there are good explanations for why some of them are as common as they are, even though they hugely diminish the chances of the sufferer passing on his or her genes. That is the comparison I am making because, when all is said and done, male homosexuals do nowadays have a lot fewer children than most heterosexual men. That is the puzzle. If there is a gay gene, why is it so common? Why did it not become extinct long ago, unable to reach, or least constrained from reaching, the next generation? These are exactly the same questions we need to ask about a serious genetic disease. The fact that homosexuality is not a disease doesn't matter so long as it reduces the chances of the gene being passed on. If having brown eyes meant you didn't have children, no-one would have brown eyes any longer.

In medical genetics there are only a handful of available explanations for the persistence of an inherited disease. The simplest is that every new case is caused by a fresh mutation and that is indeed the reason for one of the most familiar, the type of dwarfism called achondroplasia. It is unusual for someone with achondroplasia to have children and only about 20 per cent of patients with achondroplasia inherit it from one of their parents. The other 80 per cent have achondroplasia because the same mutation keeps occurring in the germline cells of otherwise normal people. Other inherited diseases don't have anything like the high mutation rate of achondroplasia, yet are still common. The best example is the blood disease called sickle cell anaemia, where the gene affected codes for one of the protein chains that make up haemoglobin. Haemoglobin is the main component of red blood cells – it gives them their colour – and its job is to transport oxygen and carbon dioxide around the bloodstream from the lungs and back. In sickle cell anaemia the haemoglobin is unable to carry oxygen quite as well as normal. The molecules of haemoglobin clump together inside the red blood cells and this changes the shape of the cell from a disc, looking rather like a flying saucer, into a crescent – hence the name of the disease. On their journey through the body, red blood cells have to squeeze through extremely narrow blood capillaries in the tissues in order to supply them with oxygen. The smallest capillaries are even smaller than the blood cells, so the cells get squashed into sausage shapes as they squeeze through. That is no problem for the normal, flexible red cells, but the much more rigid sickle cells get stuck in the capillaries and block them up. This in turn leads to tissue death and gangrene. The cells are also liable to burst, sending the haemoglobin count right down and leading to severe anaemia. The spleen swells up to enormous size trying to cope with the task of recycling the debris from the shattered cells.

Sickle cell anaemia is a very nasty disease, and children who inherit it die very young and never have their own children. Without lots of fresh mutations happening, how come the deadly gene is still around, since it kills those who suffer from it before they can reproduce? Surely it should have been immediately eliminated? Part of the answer comes from recalling that we all have two sets of chromosomes. The sickle cell gene is on chromosome 11 and people who suffer from the disease have the gene on both copies of their chromosome 11. They have, in effect, a double dose of the mutation. Their parents both have one normal chromosome 11 and one with the sickle cell gene. They are known as *carriers* of the disease. They can carry a single gene and are not weighed down by two. With only a single dose of the sickle cell gene, their haemoglobin is good enough and their red blood cells don't sickle – unless they experience low air pressure, for example, by going up a high mountain or on a long airline flight. They don't have anaemia, they don't get ill and they do have children. So, from the sickle cell gene's point of view, it is perfectly safe being in a carrier. It does stand a chance of being passed on to the next generation.

But its long-term prospects overall are still pretty gloomy, because every time it joins up with another copy of itself, and that happens on average in one out of four children when both parents are carriers, that's the end of the road. It isn't going any further because the child who carries both copies is going to die. Over time a gene like this will gradually disappear from the population. So that doesn't explain why sickle cell anaemia is so very common. And common it certainly is. In parts of Africa, a hundred thousand children are born with the disease every year. The explanation is that the sickle cell carriers are more resistant to malaria. Malaria is caused by a tiny parasite, carried by mosquitoes, which spends part of its complicated life cycle in our red blood cells. For reasons which even now are unclear, the malaria parasite can't get into the red blood cells of sickle cell carriers anywhere near as easily as it can enter the cells of people with two normal number 11 chromosomes. That gives the sickle cell carriers a huge survival advantage in West Africa, where malaria is endemic.

This is great news for the sickle cell gene, which can spread through the population in carriers who, because of the increased malaria resistance, are more likely to survive and have children than people without a copy of the sickle cell gene to protect them. On average, half the children of carriers will also be carriers. This compensating advantage to the sickle cell gene is enough to offset the fatal drawback of being eliminated when combined in double dose in the sufferers. You can see this works because, when malaria is eliminated from an area or the people move elsewhere, the gene – now denied its advantage – becomes gradually rarer and rarer. For example, the ancestors of many African Americans came from West Africa and carried the sickle cell gene with them to the New World. Sickle cell anaemia is, unfortunately, still relatively common in African Americans but, in the absence of malaria, the gene is already becoming rarer and will continue to do so.

Those are the two main reasons why a serious inherited disease can be common: a high mutation rate or a compensating advantage. I said at the beginning of the chapter that male homosexuality puzzled me because it resembled a serious genetic disease. But is male homosexuality genetic at all? The classic way of exploring whether or not a characteristic has a genetic component is to monitor its occurrence in twins. About one in ninety pregnancies results in twins. A third are identical twins and the remaining two-thirds are non-identical. Identical twins both develop from the same fertilized egg; thus they both inherit exactly the same sets of genes from their parents and, barring subsequent mutations, are genetically absolutely identical. Non-identical twins develop from two separate fertilized eggs and do not inherit the same genes from their parents. On average, though, they will still have half their genes in common. In this respect they are just the same as siblings and, just like siblings, they can be either both the same sex or one of each. How is this helpful? Imagine you had no idea whether sickle cell anaemia was genetic or not. If you were able to find pairs of identical and non-identical twins, one of which in each case had sickle cell anaemia, you could get a clue by seeing how often the other twin also had the disease. I'll tell you what would happen. In identical twins, where one twin had sickle cell anaemia the other one would also have it – 100 per cent of the time. That's an indication of a genetic influence, but it's still not proof because, for example, both twins might have been affected by conditions in the womb or a shared environment in early childhood. The way to iron these environmental influences out, as far as possible, is to compare what happens in identical twins with what happens in pairs of non-identical twins. Like identical twins, non-identical twins share the same womb, are both born more or less at the same time and, if they are brought up together, will usually share much the same environment. However, they have only half their genes in common.

If a characteristic is entirely genetic, like sickle cell anaemia, then pairs of identical twins will share it every time. If one twin has it, so will the other one. They must, because they have all the same genes. If it is entirely genetic, non-identical twins will share the same feature less often than identical twins because they have fewer genes in common. On the other hand, where a characteristic has no genetic component at all and is determined completely by the environment, there will be no difference in how often it occurs in the two types of twins. For instance, there is as much chance of identical twins being struck by lightning as there is of the same thing happening to non-identical twins. Most human characteristics are a result of a mixed influence of both genes and environment. This is the familiar nature versus nurture debate which divides opinion whenever it surfaces, as if a characteristic – intelligence, criminal behaviour, musical ability, athletic prowess, you name it – must be attributable to either one cause or the other. Of course, the answer is always that genetics *and* the environment, nature *and* nurture, both make their contribution. The only thing worth debating is the relative influence of the two forces. Studying identical and non-identical twins can give some idea of their relative importance in forming any character, and over the last eighty years and more a great deal of research has examined every conceivable characteristic in twins to try to estimate the extent to which nature and nurture contribute. The sign of a genetic component is when identical twins both show the characteristic more often than do non-identical twins. Most work has concentrated, not surprisingly, on medical conditions, and that is where the most reliable data are to be found. Take coronary heart disease as an example. If an identical twin has coronary heart disease, then the likelihood of the other one developing coronary heart disease is 46 per cent. This figure of 46 per cent is known as the *concordance rate* for coronary heart disease in identical twins. The concordance rate in non-identical twins is much lower, only 12 per cent, and this is indicative of a fairly strong genetic influence in coronary heart disease. For diabetes the concordance rate for identical twins is 56 per cent and for non-identical twins only 12 per cent – again indicating a substantial genetic contribution. The genetic contribution to psychiatric illness is apparent from figures for schizophrenia, for example, where identical twin concordance is 45 per cent against 12 per cent for non-identical twins, and also for manic depression or bipolar disorder (identical twins 70 per cent, non-identical 15 per cent). There are huge arguments about whether these figures are reliable, depending as they do on accurate definition and diagnosis, which is particularly difficult in the psychiatric illnesses.

These arguments intensify when it comes to questions of intellectual performance and behaviour, but there are data to mull over. For example, in an old study from Germany, twins of both sorts were compared even down to their school grades for different subjects. The concordance for identical twins was higher in all subjects – except English! Make of that what you will. My own view is that you can go on debating the meaning and accuracy of these sorts of twin studies until you are blue in the face. They might be useful as an indicator of a significant genetic component in a particular condition or tendency, but that is all. Their value lies in being a guide for further research. If you want to find the genes for a particular characteristic, it's as well to have an indication that there are genes to find before spending time or money on what, in any event, often turns out to be a wild goose chase.

The concordance rates for male homosexuality certainly do indicate some degree of genetic influence. One study in the 1950s showed a concordance of almost 100 per cent for homosexual behaviour in identical twins and of only about 20 per cent in non-identical twins. The hint of a genetic component that emerged from these reports, flawed though they undoubtedly were in ways we need not go into, was sufficient to encourage Dean Hamer and his colleagues from the National Institutes of Health in Washington DC to take a closer look at the genetics of male homosexuality. Hamer carried out his own twin study which, while arriving at quite different figures, still showed the same trend. His concordance rate for identical twins was 57 per cent and less than half that (25 per cent) for non-identical twins. Encouraged by this,

Hamer began to search for the gene, or genes, involved in male homosexuality and in July 1993 published his findings in the leading American journal *Science*. In this paper he claimed to have found a region of the X-chromosome on which a gene predisposing to male homosexuality was located. Predictably, within hours of publication, the news of the discovery of 'the Gay Gene' was flashed round the world. What Hamer had done was to recruit gay volunteers either at the local AIDS clinics in Washington or through advertisements in gay magazines. He took DNA samples from these men and from as many of their relatives as were prepared to take part. From drawing out their family trees, Hamer noticed that a lot of the gay men also had gay uncles, but only on their mothers' side. This pattern of inheritance is reminiscent of characteristics whose genes are carried on the X-chromosome, among them the blood-clotting disease haemophilia and the common type of red–green colour blindness.

Only men get haemophilia and only men are colour-blind. Because the genes for both haemophilia and colour-blindness are carried on the X-chromosome, and recalling that men have only one X-chromosome while women have two, it is only men who are affected because the mutated gene on their X-chromosome cannot be masked by a normal copy of the gene. In women, on the other hand, even if one of their X-chromosomes carries the mutation, they have another X-chromosome which carries a normal gene and which will over-ride the mutated copy. The mother of a haemophiliac or colour-blind son is a carrier with one normal and one mutant copy of the gene. The son who gets her mutated X-chromosome will get haemophilia or colour-blindness as the case may be.

Working on the hypothesis that male homosexuality might be inherited in a similar manner, Hamer and his colleagues then tested the volunteers and their relatives to see whether or not in each family tree the gay men had all inherited the same X-chromosome, the rationale being that, if they had, this was proof that a gay gene lay somewhere on that chromosome. To make the search even more specific, he managed to follow different segments of the X-chromosome through the family pedigrees, so that if a particular part of the chromosome was shared between the gay men, this would pinpoint approximately where on the X-chromosome the gay gene might be found.

The most remarkable results from his study were from the forty pairs of gay brothers that he had recruited through magazines. Normally two brothers could expect to inherit the same X-chromosome from their mother half of the time and a different X-chromosome the other half of the time. That's because it's entirely random which of her two X-chromosomes ends up in each egg. So, if there was nothing in it, roughly twenty out of forty pairs of homosexual brothers would share the same X-chromosome and the other twenty would have a different X-chromosome. But, instead, the gay brothers shared the same X-chromosome in thirty-three out of the forty cases – far more than would be expected by chance. This was a strong result, and though it did not prove the existence of a gay gene on the X-chromosome, it made a very convincing case for it. Because he had segmented the chromosome, Hamer could also tell where on the X-chromosome the gene lay. The segment most commonly shared between brothers lay very near the tip of the long arm – by coincidence, not far from the gene for haemophilia.

The publication caused an outcry, as I am sure many of you will remember. In the perennial debate about whether homosexuality is a biological or cultural phenomenon, the apparent proof that a gene existed immediately jolted the pendulum in the direction of biology. Many gay men, who had felt either guilt or confusion about their sexual orientation, took comfort from the news that it was a gene rather than themselves that was responsible for their homosexuality. Others complained that research into the biology of male homosexuality was basically unethical and should be banned. At the other extreme, homophobes declared that a 'cure' for homosexuality through gene therapy was just round the corner. Scientific outrage was much more muted, confined largely to a predictable technical attack on the statistics used by Hamer, and the publication was followed by a chorus of algebraic disapproval. In my experience, these attacks on experimentalists' work by the guardians of statistical integrity usually come to nothing. Either the work is independently confirmed or it is not. No doubt stung by this assault, Hamer repeated his experiments with a new batch of volunteers and found similar results, though not so striking as the first. In this later study, twenty-two out of thirty-two pairs of homosexual brothers both shared the same segment of the X-chromosome. This was not as impressive a result as the original thirty-three out of forty, but is still significantly different from a random distribution. Eventually Neil Risch, the author of the original mathematical critique, decided to get his own data, and these were published in 1999. In Risch's survey of homosexual brothers, only twenty out of forty-six pairs shared the same segment of the X-chromosome in which Hamer had originally located the gay gene. That, sadly for Hamer, was not statistically different from a random distribution. That is the only large-scale attempt at completely independent replication of Hamer's original findings of which I am aware – and, in a revealing footnote, the authors declare that it was paid for out of their own pockets.

Are we any further forward? Was the whole concept of a gay gene just too disturbing and dangerous to research? We are left hanging. One set of results says there is a predisposing gene for male homosexuality, another says there isn't. The twin studies certainly suggested a substantial genetic component, and Hamer's family trees bore this out, showing gay men in several generations all connected through their mothers. It was that connection which first led Hamer, quite reasonably, to concentrate his search for the gay gene on the X-chromosome. But, as we saw a few pages back, the huge disadvantage that any gay gene imposes on itself by severely limiting its ability to move through to the next generation would have to be compensated for by an absolutely massive counterbalancing advantage to the carriers – the mothers and sisters of the homosexual men. Without that huge advantage any major gay gene would be dead in the water.

Is it possible that the female carriers of a gay gene could have a massive selective advantage over other women? It is actually surprisingly difficult to pin down what the selective advantage of carriers actually is. It took decades to prove that it was their resistance to malaria which gave carriers of sickle cell anaemia the edge over their compatriots without the mutant gene. But there is tremendous uncertainty about the advantage enjoyed by the carriers of the most common inherited diseases in people of European ancestry – cystic fibrosis and haemochromatosis. All sorts of theories abound. One in twenty Europeans are carriers of cystic fibrosis and a staggering one in six carry one copy of the haemochromatosis gene. Yet in double doses these genes are dangerous. So why are the genes still going? Until very recently, most cystic fibrosis patients died by the age of twenty because the faulty gene prevented them from clearing mucus from their lungs. After numerous bouts of recurrent lung infection, cystic fibrosis patients finally succumbed to respiratory failure, usually in their teens. Haemochromatosis is a less serious disease but nonetheless debilitating as the mutation disrupts the body's mechanism for disposing of iron and the metal builds up in the tissues, particularly the liver. Could the cystic fibrosis carriers have been resistant to an infectious disease that afflicted our ancestors, like cholera or diphtheria? Could carriers of the haemochromatosis gene have been better at getting the most from their iron-deficient diet in medieval times? It sounds possible, even reasonable, but there is no proof. The selective agent, whatever it was, may be long gone; we may never know.

Since we do not know for certain what advantage to carriers encouraged the spread of the cystic fibrosis and haemochromatosis genes, it is still extremely difficult to begin to imagine what possible advantage there might have been to the carriers of the gay gene which was sufficient not only to prevent its rapid extinction but to encourage its spread so widely. It's hard to imagine that carriers of the gay gene would have managed something spectacular like surviving the Black Death, but it is that sort of level of protection that is required – not just a slight increase in fertility. No, I think there has to be another explanation. Recalling the plight of the crippled Taran sperm, I began to wonder if the genetic basis for male homosexuality might have nothing to do with the X-chromosome at all, or any other chromosome, come to that.

I went back to the library to look at Hamer's original paper in *Science* and at the family trees of the gay men that he had drawn out. I could see very easily how he and his colleagues had tracked the inheritance of male homosexuality through the gay men's mothers and why this pattern had drawn his attention to the X-chromosome as the likely location of the gene. Three of the four large pedigrees had all the hallmarks of that kind of inheritance and the gay relatives were all connected through exclusively maternal links. They could so easily have been pedigrees not of male homosexuality but of haemophilia, the classic among inherited diseases of the X-chromosome, which spread its grim tentacles through the royal families of Europe in the nineteenth and early twentieth centuries. But one vital indicator of an X-chromosome pattern was missing, though for very understandable reasons. In haemophilia or colour-blindness the gene is located on the X-chromosome, as we have seen already. When they have children men, with one X- and one Y-chromosome, pass their Y-chromosomes to their sons and their X-chromosomes to their daughters. In a haemophiliac or colour-blind man, the X-chromosome, with the faulty gene, passes to his daughter and not to his son. His son receives his single copy of the X-chromosome from his mother. A son cannot possibly inherit haemophilia or colour-blindness from his own father. An X-chromosome disease will never be passed from father to son, and any instance of this in a family tree imme-

diately rules out the X-chromosome as the location of the gene. It
has to be somewhere else. In Hamer's pedigrees there were indeed
no instances of gay fathers with gay sons – but that was hardly a
surprise, since the gay men didn't have any children. Although
the pedigrees appeared to satisfy that particular requirement for
X-chromosome involvement, it was rather a case of not taking the
test rather than passing it. If fathers don't have children, you cannot know whether their sons are gay or straight. So the family trees
only pointed a finger at the X-chromosome; they did not prove the
gene lay there.

Could the same pedigrees be compatible with an inheritance
influenced not by the X-chromosome, but by mDNA? Could this
be another example of male disablement in the same league as
Tara's sperm? Looking at these pedigrees spread out on the library
desk in front of me, I began to trace the course mDNA would have
taken through the generations. The symbols and the lines that
connected them on the page in front of me began to blur as I
drifted off into a light daytime trance. It was warm in the library
and I had been looking at scientific journals all day. Outside the
weather was beautiful and my seat was near to a high window
overlooking the green lawn that lies in front of the University
Museum. A track of concrete dinosaur footprints had been laid
across the ground and mothers with their small children, too young
for school, lay close by and played in the sun. A small boy ran in a
broad circle, his arms outstretched like a plane or a bird or maybe
a pterodactyl – then back to his mother, who hugged him close to
her breast. This scene is repeated millions of times in millions of
different places every day all over the world, and must have been
for thousands upon thousands of generations. Here was the bond
of love and nurture which stretches back generation upon generation into the deep past; the bond which I had already followed
around the world using the piece of DNA that defines the essence
of femininity and continuity – mitochondrial DNA.

I awoke from my daydream and focused sharply on the page in

front of me. I don't know why, but when I looked back at the pedigrees, the answer leaped out. Of course, male homosexuality had nothing to do with the X-chromosome but everything to do with mitochondria. Everybody gets their mitochondria from their own mothers, but only daughters pass it on. Mitochondrial DNA might be a symbol of femininity, but it still carries genes with the blind ambition of getting through to the next generation and beyond. The mother playing outside with her young son obviously loves him – but her mDNA doesn't. From its point of view it would have been much better if he had never been born, never even been conceived, so that she could concentrate on having daughters. My mind was racing. What was forcing her to have sons? Her husband's Y-chromosome – nothing else. And what stood to gain from her having sons rather than daughters? Same answer: her husband's Y-chromosome. And what got passed on to the next generation of his sons? His Y-chromosome. Her mitochondria would do much better if she could eliminate all her male foetuses, just as the lady from Alsace and her family had managed to do.

But if she failed to kill her sons in the womb, failed to crush the Y-chromosome during its most vulnerable nine months when she carried it within her own body, then she would see to it that it got no further. She would turn her son into a homosexual. The effect would be just like Tara's poisoned kiss which disabled her sons' sperm. I could see immediately that this hypothesis solved the major theoretical obstacle to the 'gay gene' paradox in a genetic sense – the puzzle of how such a gene could survive and not be eliminated by its failure to be passed on through gay men. That vanished at once, because if the genetic element were associated with mitochondrial DNA, with the cytoplasm, it wouldn't get passed on by men anyway. It is inherited entirely maternally from mother to daughter. It really felt as if a huge boulder – having to explain how any gay gene had survived – was suddenly rolled out of the way.

I looked back at the pedigrees and saw that it would work. A mitochondrial inheritance was just as possible as an X-chromosome association. A mother passes her mitochondrial DNA on to all her children and, of course, there were plenty of examples of men whose brothers were gay but who were not gay themselves, even though they had the same mitochondrial DNA. But I didn't see that as a problem. I never imagined the mechanism for making a son gay was actually encoded by the mitochondrial DNA itself. That just supplied the motivation. Perhaps mothers whose sons became gay had just not managed to eliminate them while they were in the womb. There was no necessity, in my rapidly forming theory, to disable all her gay sons. It was a battle with an uncertain outcome in every new pregnancy. My mind was racing now. If gay sons were the victims of failed attempted intra-uterine elimination, did their mothers also have a record of successful prenatal homicide? Had they managed to kill sons before? I looked again at the family trees. Did the gay men have more sisters than brothers? Not particularly. In the families of the gay men there were roughly the same numbers of brothers and sisters. But when I looked back a generation to see whether the mothers themselves had more brothers than sisters, there were far more girls than boys. I found out later that this was generally true. In a survey of nearly five hundred gay men, their mothers had a total of 209 sisters but only 132 brothers. Of course, they ought to have had roughly equal numbers of brothers and sisters. These gay men had far more aunts than uncles. So what happened to the missing seventy-seven brothers? Had they been killed while in the womb? Had these mothers been even more successful at eliminating the male embryos and their Y-chromosomes than their daughters, who could only neutralize their sons by steering them towards homosexuality?

There is plenty of evidence to show that culture and environment have an important influence on sexual orientation. But there are also hints of a few biological mechanisms which would give the mother at least the opportunity to influence the sexual orientation of her sons while they were still in the womb. Let me preface my descriptions by saying that there is certainly no general agreement among scientists on any of these mechanisms – in fact, quite the reverse. The scientific literature on the biological basis for sexual orientation is a battleground of claim and counterclaim. With that proviso, here are some of the possibilities. For the most part they revolve around the notion that, just as male anatomy develops in the foetus under the direction of testosterone away from a feminine developmental pathway, so development of the male brain is a diversion from an otherwise female plan. Under this scheme male homosexuality is explained by a hitch in the transition to the male pattern. The anatomy of men's and women's brains is surprisingly similar, even though they act and think so differently, and only after a lot of detailed comparisons were any consistent differences found between the two. One of them lies within the hypothalamus, and its detailed description is 'the central subdivision of the bed nucleus of the stria terminalis' – or BST for short. It would take another chapter to explain just what this is, but all we need to know here is that the BST is two and a half times bigger in males than females, that it has plenty of sex hormone receptors and that it is wired into another brain structure, the small, almond-shaped amygdala. The amygdala is like a crossroads in the brain: the hub of an interconnecting network of neurological pathways and the seat of many of our emotions. The clue to the BST's association with gender identity and sexual orientation came when a team of Dutch scientists from Amsterdam conducted post-mortem examinations of the brains of six male-to-female transsexuals, men who had from childhood onwards had a strong feeling that they had been born the wrong sex. The Dutch team found that the BSTs of these men were much more similar in size and structure to those found in the typical female brain than to those of a man's brain. These men were transsexual rather than homosexual, and the Dutch team are continuing their work to see if sexual orientation as well as gender identity can be pinpointed to the same part of the brain.

The discovery of a structure that had such an impact on gender identity and sexuality, and was formed early on in the developing brain of the foetus, tied in with experiments that had been done some years before with rats. Homosexual behaviour in rats could be induced by artificially lowering their testosterone level while in the womb, but only if this was done at a critical time for brain development. These and similar experiments led to the theory that sexual orientation was decided at some key point in the development of the foetal brain while it was under the influence of sex hormones, both those circulating in the mother and those being produced by the foetus itself.

One other strange observation also suggested that sexual orientation is decided in the womb. Did you know that foetuses suck their thumbs? Everyone knows that children do, but I didn't know foetuses did as well. But they do. Using ultrasound scans, scientists discovered that 92 per cent of them suck their right thumb, which is close to the percentage of adults that are right-handed. Even at only ten weeks, foetuses move their right arm three times more often than their left, and a ten-year follow-up study showed that the hand they used as a foetus was also the one they preferred as a child, and presumably will prefer as an adult. The connection between handedness and sexual orientation is this. It turns out, from studies done over many years, that homosexuals are far more likely to be left- than right-handed. Since handedness is an early neurological development it follows that sexual orientation probably is as well.

The final piece of evidence, if you can call it that, also has something to do with hands. Have a look at your fingers – on your left or right hand – it doesn't matter which. Open your hand flat and look at your index finger, the one next to your thumb, and compare it to your ring finger, the one next to your little finger. Is your ring finger noticeably longer than your index finger, or are their tips more or less in line? If you are a woman, the chances are that the two fingers are almost the same length. In men, the difference in their lengths is much more noticeable, with the ring finger always longer than the index finger. What has that got to do with sexual orientation? In 1999 a team of Californian researchers went round public street fairs in San Francisco and asked 720 adults about their sexual orientation, then measured their fingers. When they sat down to analyse the results, they first discovered that the sex difference in finger length was greater on the right hand than on the left. When they compared the ratios with sexual orientation, they discovered that the finger-length ratio of homosexual women was much more like that of the men, with a relatively shorter index finger, than it was like that of the heterosexual women. Among the homosexual men, though, the finger-length ratios were just the same as among heterosexual men.

Fingers are formed early on in the foetus and their relative lengths are influenced by androgens, sex hormones like testosterone. The Californian researchers suggested on the basis of their results that homosexual women had been exposed to more androgens than heterosexual women while they were in the womb. Since they didn't find that the finger-length ratios in the gay men were any different from those of straight men, they could not conclude that male homosexuality was associated with prenatal exposure to lower levels of androgens. However, they did confirm one thing about male homosexuals that had been noticed before. They more often had older brothers than heterosexual men.

The research on finger lengths, handedness and the brains of transsexuals all pointed to individuals' sexual orientation and gender identity being influenced very early on, during the time they are growing in their mother's womb. In my idea of male homo-sexuality being a way for a mother's mDNA to prevail at the expense of her son's Y-chromosomes, her best chance of engineering that is when the growing foetus is in the womb, and that is what all these pieces of research indicated was going on. The evidence on elder brothers also suggested another way she might do the same thing. The cells of males, including when they are growing in the womb, have on their surface a molecule called the H-Y antigen, with its gene on the Y-chromosome. Rather like a blood group protein, or one of the tissue-typing molecules that have to be matched before an organ transplant, H-Y can be recognized as foreign by anyone who doesn't have it. And, being female, mothers don't have H-Y. When a woman is carrying her first male child, a few cells from the foetus get into her circulation where they are recognized as foreign because of H-Y. Nothing happens to the child, but the mother will begin to make antibodies. She is, in effect, being immunized against males, and the next time she is pregnant with a male foetus she can try to reject it, just as we all defeat infectious diseases when we have had a jab.

Ray Blanchard and his colleagues from the University of Toronto have knitted all these observations into a hypothesis which sees the antibodies cross the placental barrier and find their way into the brain of the male foetus. When that happens, according to Blanchard, the antibodies interfere with and partially block the sexual orientation centres of the foetal brain, presumably including the BST, from developing along the normal route, with the result that the son will be attracted to men rather than women. This affects only second and subsequent male foetuses, but with each one the intensity of the immune effect increases as the mother is re-immunized with each pregnancy. This theory is used to explain why the chances of a man being gay increase the more older brothers he has. According to Blanchard, the probability goes up from 2.6 per cent for a boy with just one older brother to 6 per cent for a boy with four older brothers. Sad to say, when he published his theory Blanchard was asked by reporters on several occasions whether he thought it might incite homophobic parents to abort male foetuses if they had already had two or three sons.

I am only too well aware that my theory is inadequate as a complete explanation for male homosexuality, and it isn't meant to be that. I am just happy that the headache that has dogged me for years about the virtual impossibility of an orthodox gene for male homosexuality surviving rapid extinction has now stopped throbbing, even though it has not completely cleared. That the homosexual man and his Y-chromosome are casualties in the genetically embedded war between the sexes makes much more sense. But are the motives purely those of revenge? Could a mother's mDNA actually have anything to gain from having a gay son? For some time I couldn't see what it could possibly be. Then, much later, I realized an answer lay in the beehive in the museum just beyond the grass where I had seen the mother and her son playing. Could a gay son possibly be doing for his mother what the sterile workers in the hive were doing for their queen bee? Could a gay son be helping his mother to bring up his own sisters? That would be a direct benefit to the mother's mDNA. Any such small advantage would be very useful indeed and mDNAs with that ability would do very well, irrespective of the fact that they had made all their sons sterile. That would elevate male homosexuality to a true piece of genetic altruism. It is a subtle plan by mDNA, not only to get rid of Y-chromosomes but to help itself at the same time.

24

GAIA'S REVENGE

We have got to the point in the story where we can look to the future. We have seen the fundamental reason for sex in us and in most other living species. We have the explanation for the universal division into two separate sexes, one the guardian of the egg, the other the broadcaster of sperm. We have seen how this very fundamental division, and the different genetic interests it confers, lies behind the often very distinct behaviour patterns of the two sexes that we see in ourselves and other animals. We have also seen how the two principals, each of whose genetic future is linked to just one sex and not the other, are reporters, warriors and perhaps even instigators of the enduring conflict, even if we do not always know how they put their battle plans into effect. You also have my own view that sexual selection, seizing on the new factors of wealth, property and ownership that emerged with the invention of agriculture, has transformed our world beyond recognition. In no time at all, and operating with essentially the same gene pool as our ancestors, we have changed from small-scale hunter–gatherers, reliant on wild food and the earth around us, into a global and urbanized species that has largely severed its links with nature.

Of course, not all of these changes have been bad, and no-one would seriously advocate a return to the hard life our ancestors endured twenty thousand or more years ago. But there is no denying that sexual selection, acting through wealth and power, has severely disturbed the balance between the two sexes and created the patriarchal social structures where men seize and retain control. The major genetic beneficiary of this process has been the Y-chromosome, or at least those Y-chromosomes that have managed to hitch themselves to wealth and power and have been able to multiply beyond all expectation. They have been helped on their way through future generations by the common rules of paternal inheritance whereby wealth, property, title and name are usually passed from father to son.

Looking ahead, is there any limiting force acting to inhibit the effectiveness of this sexual selection? There are natural limits on how big a peacock's tail can become before he is unable to fly. Such a bird might have the pick of the females, but he would also be picked off himself, by a predator, if he couldn't fly up to roost at night. His tail genes, magnificent though they no doubt are, do not get passed on. Equally, the male elephant seal's bulky frame is only any help if he can actually get up the beach to the females. A real whopper who is stuck in the surf because he is too fat to support his own weight out of the water isn't going to breed – so his genes are finished. But when it comes to wealth and power, it is hard to see any natural limits such as those that restrain the peacock and the elephant seal. Indeed, quite the opposite appears to be the case: among humans, the rich generally get richer while the poor get poorer. I have blamed the blind greed of our species on sexual selection, greed which is slowly but surely destroying our planet in ways we all know. Therein lies the greatest danger. It is very hard to see any salvation in the normal rules of evolution acting, as we have seen, for the good of genes and not the species. We might despair at the destruction of the natural world, even as it destroys our own species, but that will not stop the process. Genes are blind and have no concept of the future.

It is always extremely difficult to predict what lies ahead, but I find it hard to imagine the world in a thousand or ten thousand, let alone a hundred thousand, years from now. These are comparatively trivial time spans when we look back into the past, well within the recollection of the genes we carry. But looking ahead, there has to be a distinct possibility that, at the rate we are going, we will so damage the world that it can no longer support us. Without labouring the point, already within my lifetime we have been on the brink of a nuclear war, in the 1962 Cuban missile crisis, and, even as I write, a war is under way in the Middle East. Forests are being cleared at an alarming rate, oil pollutes the beaches and acid rain falls from the skies. This is a familiar litany to all of you and, other than as a geneticist, I have no particular qualification to comment on it. I have made a case that all of this can be traced to the fundamental genetic differences between men and women and the way in which female 'choice', in its many guises, has encouraged the exaggeration of these trends. Of course, it would all be quickly reversed if women preferred to mate with men who held assets that were the antithesis of wealth and power, and if the purposely wasteful displays of Ferrari and Rolex were no longer effective. Then the runaway train of sexual selection would soon slow down. Where Eve chooses to go, Adam is bound to follow.

Setting this aside, what else might happen? Sexual selection, the basis of Adam's Curse, operates because women have eggs and men have sperm. Eggs need sperm, and Y-chromosomes need them even more. But sperm are in a bad way these days. A team of scientists from Copenhagen collected together the results of sixty-one separate surveys of sperm count from 1940 to the 1990s. The fall is so dramatic that I thought you should see it, and have reproduced their findings in figure 5.

The dashed line shows the drastic drop in the percentage of men with sperm concentrations in their semen of over 100 million per millilitre. In the 1940s 50 per cent of the men in the surveys had

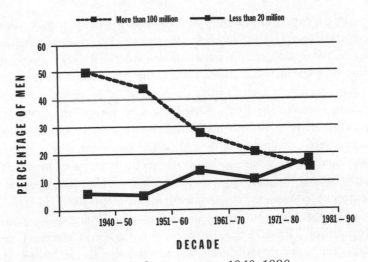

Figure 5: Sperm counts, 1940–1990

Source: E. Carlson et al., 'Evidence for decreasing sperm quality of semen during past 50 years', British Medical Journal, vol. 305, 1992, pp. 609–13

over a hundred million sperm per millilitre; that had dropped to 16 per cent of men by the 1990s. Conversely, the percentage of men with fewer than 20 million sperm per millilitre of semen climbed from 6 per cent in the 1940s to 18 per cent by the 1990s. These were all normal men with no history of infertility. Wherever you look, the sperm count is falling fast. Though most of the surveys which were included in drawing up the chart were carried out in the United States or western Europe, they were not exclusively from these regions. Low sperm counts were found in men from places as far apart as Peru, India, Libya and Nigeria. In fact, they have generally fallen so far that the lower limit for a 'normal' sperm count in infertility centres has had to be revised downwards from 60 million to 20 million per millilitre. Low sperm counts are a major factor in male infertility. This is serious. Until this careful summary was made, no-one really took any notice of the danger signs, because most of the previous reports of falling sperm counts had come from studies on selected groups of men recruited through infertility clinics, where you might expect to find men with low sperm counts anyway. But after the Copenhagen study was published in 1992, research intensified to try to uncover the causes of this worrying trend. The first thing that scientists discovered is that, compared to other animals, human sperm is in pretty terrible shape anyway, even at the concentrations found in the 1930s. Between a third and a half of human sperm is anatomically abnormal, adopting an array of weird shapes that prevent it swimming in the right direction. Some people excuse these deformities as devices to outwit sperm from other men that might be after the same egg. I doubt this is true because in our closest primate cousins, the chimpanzees, whose sperm really do face a lot of competition from other males because female chimps are copulating right, left and centre, 100 per cent of the sperm is in excellent shape. Our poor performance against other animals can be judged by the fact that men produce approximately the same number of fit sperm per day as do hamsters, only a fraction of our size. Considering the damage it has wrought on the planet, through the machinations of Adam's Curse in all its guises, it is a surprise to find our sperm in such bad condition.

This poor quality also makes it very difficult to discover what is responsible for the sad state of human sperm. Such is the variability among men, and among sperm counts taken on different days, that it is hard to pin down any effect of occupation or lifestyle because such large numbers of men, usually several hundred, have to be included in a study for it to stand any chance of proving anything. Big surveys are expensive and difficult to organize, so only things that had an absolutely catastrophic effect on the sperm count were ever picked up by the early researchers working with dozens rather than hundreds of men. One of these was a pesticide called dibromochloropropane or DBCP for short, which decimated the sperm count of any man exposed to it. It was quickly banned. DBCP was so deadly for sperm that its lethal qualities were easy to spot in even a small survey. Plenty of other chemicals showed inconclusive results in surveys of similar size, leading to a certain amount of complacency in the chemical industry and the assumption that they were harmless whereas, in fact, their toxicity would have been disguised by the small scale of the studies. One other interesting discovery came out of the sperm count work in the shape of another, hitherto unknown variable. Despite our belief that we don't have a breeding season, we are clearly seasonal mammals when it comes to sperm. Counts are much higher in the winter than in the summer. The best explanation of this is that we are primed to have our children in the autumn, the best time of year for survival in ancient times when food was at its most abundant. Yet another genetic legacy from our hunter–gatherer ancestors.

When the results of larger studies, capable of detecting only modest effects on the sperm count, began to come in, some clear trends emerged. The first was the effect of testicular temperature. In most mammals, including humans, the testes are held outside the body in order to cool them down by a couple of degrees. If they were kept at normal body temperature, sperm production would fail completely; in fact, so sensitive are they to temperature that warming up the testicles has been proven as a very effective, and easily reversible, contraceptive. You don't need fancy pills; just a pair of electrically heated Y-fronts does the trick. Bakers, welders and furnace workers all suffer from high testicular temperature and low sperm counts, as do taxi-drivers and other men who spend all day sitting down without allowing a draught of cooling air to waft over the parts. Tight underwear and hot testicles do have an important effect on sperm counts, but they are not hard to reverse and not particularly sinister. Much more worrying for men are the effects of environmental pollution, particularly pesticides. Their effects are very difficult to measure, partly because of the innate variability of sperm counts already mentioned, and partly because there are so many of them. They find their way into our food, and surprisingly high levels of pesticides have built up in our own fatty tissue. Astonishingly, this includes old residues from pesticides which have now been banned, absorbed when they were still in use. These residues are still there in our own fat cells and a major concern has been their transfer en masse to new-born children when mothers mobilize their fat reserves to produce milk, particularly as this is an important time for the developing male testis.

We saw in the last chapter that the correct balance of hormones is crucial during male sexual development in the womb. This was brought home by the treatment of several million pregnant women between 1949 and 1971 with the synthetic oestrogen diethylstilbestrol which, years later, severely reduced the sperm count of sons exposed to it in the womb. There are also reports that ethinyl oestradiol, a synthetic oestrogen used in the oral contraceptive pill, is sometimes found in drinking water. Oddly enough some pesticides can mimic the sex hormones, particularly oestrogen, and are hormonally active. Soya is also a rich source of oestrogen mimics and its consumption, as an allegedly healthy substitute for meat protein, has rocketed in the past thirty years. The same is true of other chemicals with which we, and more importantly our food, come into contact every day through the universal use of plastics in the modern world. Among those with the most potentially dangerous effects are the phthalates, known from animal experiments to counteract androgens like testosterone. Phthalates are used in very many plastics and coatings from which they leach out, especially during microwave cooking. Though human exposure to phthalates is below the levels which caused problems in the animal experiments, the concentration needed to reduce human sperm counts is not known.

One extraneous source of hormones which has not received very much attention is the treatment of livestock with sex hormones to promote growth. Although this was banned in Europe in 1981, it still continues in the United States where the use of extremely potent oestrogens is routine. Even though whether they reach us in active form is disputed, exposure to even tiny amounts of these oestrogens is a cause for concern. A related issue is the hormones we get through drinking cow's milk. Unlike women, cows continue to lactate and are milked throughout their pregnancies, in the latter half of which the levels of oestrogen, and another female hormone progesterone, are extremely high. Fortunately, or so it is claimed, the oestrogen is destroyed during the formulation of powdered milk for babies, though it is not clear how this happens.

These are certainly all worrying trends for sperm and for male infertility, and deserve to get more attention. But what delicious irony that it is male fertility that suffers first from the poisoning of the planet. It is almost as if Gaia, now fully awake and aware of the terrible effects of Adam's Curse, is targeting her comeback in just the right place. Gaia's revenge is hitting men where it hurts most.

25

LIFTING THE CURSE

During the course of the book we have seen how the main bulk of our genes, on the nuclear chromosomes, are practised at backing whichever sex suits their purpose at the time. They have no loyalty to either sex. In contrast to this fickle behaviour, our two principal characters, the mitochondria and the Y-chromosomes, are totally committed to one or the other. Their survival depends on it. They have this much in common, but their natures are very different. For a start, mitochondria are not intimately involved in deciding sex in the way that Y-chromosomes most definitely are. Though I think they are perfectly capable of *influencing* the sex and perhaps even the sexual orientation of children, they are evidently not the principal initiating factor. After all, both men and women have mitochondria. All cells, both male and female, need mitochondria, and although the nuclear chromosomes have done their best to capture mitochondrial genes over the course of evolution, they have learned to tolerate each other. Mitochondria are here to stay.

The Y-chromosome, on the other hand, is in a mess. While mitochondrial DNA is a model of slimmed-down efficiency, the Y-chromosome is a genetic ruin, a wasteland littered with molecular wreckage. There are more active genes in the sixteen and a half thousand bases of mitochondrial DNA than in the sixty million bases of a Y-chromosome. Why is it in such a state? To answer that we need to look back towards its origins. Originally the ancestral Y-chromosome was a perfectly respectable chromosome, just like the others, with a collection of genes doing all sorts of useful things – but its fate was sealed when it took on the mantle of deciding sex. This probably happened in the very early ancestors of the mammals, perhaps 200–300 million years ago. A mutation on one of those ancestral chromosomes suddenly, and quite by chance, enabled it to switch on the pathway to male development. That doesn't mean to say that before this mutation occurred there were no males, but that they were 'switched on' by some other means. This could have been in one of the many different ways we have already encountered, like incubation temperature of the alligator's egg or the social hierarchy of the wrasse. Or it may have been chromosomal, with a gene somewhere else doing the switching. The new mutation might have been in one of the genes further down the chain of command, which normally had to wait to be activated by the original switch. Nobody knows precisely what the mutation was, and it is not important. What matters are the events which this chance event set in train. As soon as this gene took over the decision-making, the chromosome which carried it was doomed.

For reasons that not all scientists can agree on, and which need not trouble us here, a newly appointed sex chromosome is denied the advantages of recombination with its former partner which in mammals, including humans, was almost certainly the X-chromosome. We can tell this because there are still a few genes on the human Y-chromosome which have recognisable counterparts on the X-chromosome with similar DNA sequences, pointing to a distant common ancestry of the two chromosomes. As a memory of this once happy marriage, the X- and Y-chromosomes still embrace, if only very lightly, at their tips when cells divide. However, between these fleeting contacts, the rest of the Y-chromosome has been shunned by its former partner and prevented from enjoying the benefits of sexual recombination by the very system it maintains – sex itself.

Once a chromosome has been denied the opportunity to recombine with its partner this limits its capacity to repair the damage inflicted by mutation. Sexual recombination has a healing effect, allowing damaged genes to be rescuscitated by their healthy companions on the undamaged chromosome during the 'final embrace' before they go their separate ways to sperm or egg. Chromosomes denied this nursing care get sicker and sicker. Mutations, almost all of which are inevitably harmful, silence genes one after the other. The human Y-chromosome is a graveyard of rotting genes, whose corpses are nonetheless still sufficiently similar to active counterparts on the X-chromosome to be recognizable, but whose festering remains contain the evidence of their own demise – here a few bases cut from a key section; there a spelling change that makes a nonsense of a once vital instruction. Without any capacity for repair, the mutations keep on accumulating. Like the face of the moon, still pitted by craters from all the meteors that have ever fallen onto its surface, Y-chromosomes cannot heal their own scars.

This was the widely held view until very recently, when, in June 2003, a team of US scientists led by David Page announced in *Nature* the discovery of a new and completely unexpected mechanism by which the Y-chromosome might indeed be able to repair itself. Page, whom we met earlier during the hunt for SRY, accomplished the technically demanding task of sequencing the Y-chromosome of a single man and discovered within it eight strange islands of DNA. These islands, scattered through the otherwise ruined landscape of dead and dying genes, are written as immensely long palindromes where the sequence reads the same forwards as it does backwards: a genetic version of 'A man, a plan, a canal – Panama' but much, much longer. These palindromes, which can stretch for hundreds of thousands of bases, also contain genes which are active in testis cells and therefore probably have something to do with sperm production. Within the DNA palindromes, these genes are often found as absolutely identical pairs, and Page reasoned that this degree of similarity could only have been maintained if the genes were actually in communication with one another. This is indeed recombination of a sort, where the two copies in a pair meet as the DNA palindrome bends in the middle and the ends line up with each other. As this happens, the two copies get a chance to compare sequences and, by a process called *gene conversion*, repair any damage in one copy. However, this kind of internal recombination is not without its dangers. There is no guarantee that the gene conversion will repair a damaged copy. The essential ignorance of DNA makes it equally likely that the good copy will be spoiled instead. Also, the palindromes themselves are vulnerable to deletion by the very process of internal recombination that allows gene conversion to occur. If this happens, entire chunks of Y-chromosome will be lost, often taking vital genes with them. As we will see, deletions of this type are a common cause of male infertility.

The ability to recombine internally, deduced from the study of one man's Y-chromosome, was totally unexpected. It is certainly recombination of a sort, but a far cry from the vivacious gatherings enjoyed by the other nuclear chromosomes brought together by sex, where there is a chance to change partners at every generation. In comparison to these full-blown sexual liaisons, this is a lonely shuffle where each gene dances with a mirror image of itself and engages in a bit of mutual grooming. There is no chance for new gene combinations to arise, no protection against the onslaught of parasites, no contact with the outside world. It remains to be seen whether this sad gavotte has slowed down the pace of decay, or accelerated it. The Y-chromosome is as lonely as ever, but now we know it talks to itself as it spirals towards oblivion.

To make matters worse, the Y-chromosome is hit by mutations far more than any other chromosome. The reason for this extra insult is that Y-chromosomes must spend their entire lives, for generation after generation, in the cells of men. The cells that hold the Y-chromosomes ready for the next generation are in the human testis, and that is a very uncomfortable place for a chromosome to be. Genes and chromosomes in the human testis are very vulnerable to mutation. Mutations are random events that happen when DNA is copied as cells divide. So, by a straightforward numerical logic, the more cell division there is, the more mutations their DNA will experience. And cells in the testis never stop dividing. To keep up with the massive daily output of sperm (even these days) they are never allowed to rest. The cells are so overworked that the DNA of a sixty-year-old man has already been copied a thousand times before it is popped into a sperm ready for action. Compare that to the tranquillity of the human egg. Irrespective of her age, the egg cells of a woman go through only twenty-four divisions before they are released for fertilization – so the DNA in a human egg has been copied only a couple of dozen times between one generation and the next. All the egg cell divisions in females are over and done with inside the embryo, months before a girl is born. She stores these eggs, then, years later, ripens and releases just one a month from puberty to the menopause.

Our nuclear genes, excepting the Y-chromosome, have come down to us from a mixture of maternal and paternal ancestors. This means that, on average, they have spent half their time in male ancestors and half in female ancestors; half the time cushioned in the relative calm of the ovary and the other half in the hothouse atmosphere of the testis. Our mitochondria have had the smoothest passage of all, spending their entire lives in the germline cells of a long succession of women. With only two dozen cell divisions in each generation there is very little DNA copying, cutting the risk of being hit by a mutation. But it isn't just cell division that puts DNA in danger. Mitochondria themselves are very toxic places for DNA to be when they are fuelled up. They are miniature power-plants which, fed a mixture of fuel derived from the food we eat, combine it with dissolved oxygen to produce the high-energy chemical *adenosine triphosphate* – ATP for short. The mitochondria are, literally, where we burn our food. Molecules of ATP pour out of mitochondria to other parts of the cell that need energy; when they reach their destination they are discharged, like a battery, then sent back to the mitochondria for a boost. Free radicals, particularly the negatively charged superoxide ion O_2^-, are byproducts of the mitochondrial inferno. Free radicals play havoc with DNA, not only inflicting direct damage but also making the copying process far more error-prone. When free radicals are around, the DNA mutation rate shoots up.

From all this you might expect that the DNA of the mitochondria itself is in the greatest peril – and you would be right. Mitochondrial DNA is in most danger when the furnaces are going at full blast, consuming oxygen, recharging molecules of ATP and spitting out free radicals. But this is not the mDNA that is going to be passed on to the next generation. Recent research has shown that the mitochondrial furnaces in the female germline cells, the repository for the mDNA of future generations, are closed down. These cells get their ATP without using oxygen, by converting glucose to lactic acid, and so do not need to fire up the mitochondrial furnaces. It is very inefficient in energy terms but it does mean that the vital cargo of mDNA is protected from the toxic waste of the fired-up mitochndrial furnace and the mutations they inflict. It is a quite brilliant manoeuvre, saving mDNA from itself. However, it does not spare the nuclear genes, and free radicals from blazing mitochondria do seep out and attack the DNA in the nuclear chromosomes. When the furnaces are shut down in the female germline cells, their nuclear chromosomes are also spared, but in the cells of men they bear the full brunt of the toxic assault.

Compare the peaceful environment of the female germline cell, where DNA can drift down through the generations protected from harmful mutations, to the hostile conditions in the germline cells of men. They could not be more different. Not only do the male germ cells have to keep dividing day and night to keep up the supply of sperm, their mitochondria are ablaze. Even in the sperm themselves they are burning at full capacity, furiously producing ATP to supply the rapidly beating tail that propels it on its long swim towards the egg. The more they work, the more toxins they produce and the greater the risk of mutation. But the mDNA doesn't care, because sperm mitochondria are not going to get into the egg anyway. It doesn't matter a jot to male germline mDNA if it is damaged by mutation. However, the free radicals mitochondria produce certainly do diffuse out and damage the DNA in the other nuclear chromosomes. A toxic environment coupled with rapid cell division makes the male germline cell a very DNA-unfriendly place indeed – and it shows. When mutations hit vital genes they cause genetic diseases, and these damaging changes are ten to fifteen times more likely to happen in male germline cells than in their female counterparts. At least the nuclear chromosomes can rest up in the female germline cells for a generation if they are lucky enough to find themselves in a girl, which is half the time on average. But the Y-chromosome never gets the chance of a rest. It can never enjoy the tranquillity of the egg and is instead confined for ever to the hothouse of the male germline cell, locked inside generation after generation of testicles. Battered by mutation and, by cruel irony, denied the chance to use sexual recombination to make proper repairs, it is no wonder that our Y-chromosomes are in bad shape. How long can they last?

It is easy to spot the burned-out wrecks of once-active genes lying about the ruined landscape of the Y-chromosome. It is also obvious that even the few genes that are still active are peppered with mutations, though the palindromes may offer some protection. Even the ultimate master switch, the SRY gene itself, which is not in a palindrome, has been pounded. That bruising history can be easily read by comparing the detailed sequence of human SRY genes with the same gene in mice and other animals. Generally speaking, when you make a sequence comparison between genes that do the same thing in different species, they are remarkably alike. SRY genes, in contrast, are very different. Whereas most human and mouse genes are about 90 per cent the same, the SRY genes share only 50 per cent of the same sequence. The SRY genes are changing far more quickly than their counterparts on the other chromosomes, and that alone shows the long-term effect of living in the ultra-hostile environment of the testis. However, only the SRY mutations with minimal effect can survive at all. If a mutation did cripple the SRY gene so that it could no longer switch on the path to male development, any embryo that did develop would be female. That is exactly what is found in XY females. They all have a Y-chromosome, but because the SRY gene didn't work when it should have done the embryos developed into girls. Sequencing through the SRY genes of XY females, the damage becomes clear. The gene has been blasted and the mutations have hit vital parts of the instructions. When the embryos were six weeks old, the switch tried to flick to 'on' – but nothing happened. They were denied the trip along the road to masculinity and reverted to the default sex – female. And that, of course, is the end of the road for that particular Y-chromosome. It isn't going anywhere and will disappear from the face of the earth, unable to carry on to the next generation.

As well as SRY, the few other remaining genes on the Y-chromosome are very vulnerable to mutation or deletion. From David Page's recent work we now know their precise number and location, at least on one man's chromosome. There are 27 distinct genes in total, though those in the palindromes occur in more than one copy. Twelve of these 27 genes are active in many different types of cell in the body, where they probably perform a range of basic housekeeping functions not intimately associated with being a man. Four more are active in a restricted range of tissues, like brain and prostate, while the remaining eleven genes, which include SRY, are predominately or exclusively active in the testis where they presumably govern sperm production. Since the Y-chromosome is such a wreck and the few active genes are so thinly spread, plenty

of perfectly normal men have large chunks missing without suffering any ill-effects. That is because the sections they lack don't have
any of these key sperm production genes in them. However, some
men with gaps in their Y-chromosomes do experience problems
with fertility. Very often, when their sperm is examined at a fertility clinic, there is something obviously wrong with it. Either there
are far fewer than normal, sometimes none at all, or the sperm are
even more misshapen than usual, or they are very sluggish. These
men have suffered because the chunk of DNA missing from their
Y-chromosomes does contain one or more sperm production genes.
Before David Page's complete sequence became available, that is
how the few vital Y-chromosome genes were found, and sure
enough, when they were sequenced in other infertile men, they
often bore the telltale scars of mutation.

What this tells us is rather disconcerting. The historical process
of decay which is all too evident in the wretched condition of our
Y-chromosomes is far from over. It is going on all around us. An
astonishing 7 per cent of men are either infertile or sub-fertile.
There are a whole host of causes, many associated with anatomical
damage to the urethra or the development of varicose veins in the
testis. However, in roughly half of cases the infertility has no obvious physical explanation. Among these men as many as half again,
that is, between 1 and 2 per cent of all men, are infertile because of
mutations on their Y-chromosomes. That is an astonishingly high
figure when you take into account that, by the very nature of their
effects, these mutations will not have been inherited from an infertile father. These are fresh mutations which have disabled a
Y-chromosome in the father's testis which, by pure bad luck, ended
up in the one sperm which fertilized the mother's egg and produced
an infertile son.

The human Y-chromosome is crumbling before our very eyes.
What can we expect to happen if things carry on like this? There is
no reason to think they will improve – quite the reverse, in fact.
These infertile men were very unlucky that their father's successful

sperm contained the damaged Y-chromosome. The chances are that plenty of his other sperm, that lost the race to the egg, had perfectly intact Y-chromosomes. If 1 per cent of men are infertile because of a Y-chromosome mutation it's a safe bet that *all* men produce thousands, even millions, of sperm every day whose Y-chromosomes are so damaged by mutation that they would make their sons infertile if they got to the egg first. The decay of the Y-chromosome is not just restricted to the unlucky 1 per cent, it is happening right now inside every testis in the land.

If a Y-chromosome mutation is so ruinous that it makes a man infertile, then this chromosome is not, in the normal course of events, going to get passed on to his son for that very reason. However, there is one exception when it might.

Many men have overcome their infertility with the help of a fertility treatment called ICSI, which stands for Intra-Cellular Sperm Injection. First introduced in Belgium in 1992, ICSI is an extension of the well-known procedure of in-vitro fertilization (IVF), where an egg and sperm are mixed in a test tube and the embryo which grows from the fertilized egg is re-implanted in the mother's womb. That technique, first introduced to the world by the birth of Louise Brown in 1978, has since helped an estimated seven hundred thousand couples to have their own children. In straightforward in-vitro fertilization, the eggs and sperm are perfectly normal and the infertility is usually due to a problem with getting the egg from the ovaries to the uterus, often because of a blockage in the fallopian tubes.

With ICSI, the sperm do not have to be capable of fertilizing an egg on their own. They get help. Even a completely immobile sperm that could not normally fertilize an egg even if it were put right next to one, let alone had to swim anywhere, can reach its destination. It is simply injected directly into the egg with a fine needle. Once inside, its handicap no longer matters and fertilization goes ahead as normal. Then, just as in run-of-the-mill IVF, the embryo is implanted back into the mother. What could be simpler? Infertility cured. Or is it? The danger is this. If the man's infertility is caused by a damaged Y-chromosome, then ICSI will hand this Y-chromosome on to all his sons – who will themselves be infertile for exactly the same reason as their father. If that happens, they are going to need ICSI to have children too. We have merely handed down the problem to the next generation. Of course, it is not strictly accurate to call these men infertile since they clearly are not – they can now have children. But not without help.

ICSI is a special case and likely to be available only to a few for the foreseeable future. It helps crippled Y-chromosomes, which otherwise would have been rapidly eliminated by the effects of their injuries, get through to the next generation. Does that mean that we need not be worried that the Y-chromosome is falling apart so rapidly? If mutation causes infertility in 1 per cent of men, is there any need for the rest of us to be concerned? If the bullet-ridden corpses of damaged Y-chromosomes are eliminated by infertility, what does it matter if they are being battered so hard by mutation? Surely the Y-chromosome has a long future to look forward to? I don't think we can be that confident.

Certainly, badly mutilated Y-chromosomes stand no chance, ICSI apart, of being passed on to future generations. But not all mutations are quite so drastic in their effects. They can wound a gene rather than killing it outright. The wounded Y-chromosome, just a little bit more unhealthy than before, probably will get through. Then, a few generations later, another mutation will sting it again. Not fatally, but enough to weaken it a little bit more. This is death by a thousand cuts. Unable to make long-term repairs themselves through sexual recombination, and isolated from external help, the wounded Y-chromosomes will stagger on through succeeding generations, gradually becoming weaker and weaker. One by one they will succumb to the effects of the final mutations that render the men who carry them completely sterile and only then will they disappear. Other Y-chromosomes with less serious injuries will take over the task of propagating men, but that will only be a temporary respite. They too will succumb in time to the relentless pounding of mutations. As human Y-chromosomes in general become more and more unhealthy there will be a relentless and progressive reduction in male fertility, one that cannot be reversed by cleaning up the environment.

As the degeneration of the Y-chromosome has already been going on a long time, what other signs can we see of it? One predictable outcome of the gradual elimination of sickly Y-chromosomes is that there should be less variety among those that are left behind to carry on. Each Y-chromosome death permanently takes out a potential future lineage, and though other, less seriously crippled Y-chromosomes will fill the gap, there will be one colour fewer in the kaleidoscope of Y-chromosome diversity. And we do see that. The networks I drew out in chapter 16 are full of gaps, empty nodes each once filled by a Y-chromosome that has since departed. Even the strange discrepancy between the ages of Y-chromosome Adam and mitochondrial Eve might be partly explained by the random extinctions through mutation. The less variety there is left among the living, the younger the common ancestor of those who remain appears to be.

Male infertility is on the increase. Under the microscope a high proportion of human sperm from what we would consider a normal human male are already visibly deformed. Sperm counts are falling dramatically, though there are other contributory causes as well as Y-chromosome decay. The human Y-chromosome has been decaying for a very long time and will continue to do so; and we have to expect a progressive decline in male fertility as these injuries accumulate. One by one Y-chromosomes will disappear until eventually only one remains. When that chromosome finally succumbs, men will become extinct.

'But when?' I hear you ask. Before I answer that question – before I even attempt an estimate – I must urge you not to confuse what I see as the inevitability of the process with my confidence in producing an accurate figure. So here goes. For the purpose of this estimate, I am going to assume that nothing else comes into it other than the rate at which we already know Y-chromosomes are decaying at the present time. Let us begin with the figure of 7 per cent for the proportion of infertile men, and take it that 1 per cent of all men are infertile because of a Y-chromosome mutation. These mutations must have occurred in the fathers of these men, since by definition their fathers were not infertile. So these mutations are the last straw as far as that particular Y-chromosome is concerned. It cannot be saved except by artificial means. Lots of other, less serious, mutations that decrease, but do not eliminate, male fertility will also have occurred in all men. In wartime, casualty lists always include more wounded than dead, and genes under attack from mutation are no different. For the sake of simplicity, I will also take 1 per cent as the rate at which wounding Y-chromosome mutations occur in male germline cells. When passed on to their sons, these mutations make them not clinically infertile but *less* fertile – by, let us say, again for simplicity, 10 per cent. This means that 1 per cent of men in each generation will be 10 per cent less fertile than their father. Taking these figures, and in the absence of any other influence, the fertility of the whole population will decline by 0.1 per cent (1 per cent of 10 per cent) in each generation owing to Y-chromosome decay alone. What effect does that progressive decline have in the future? I won't bore you with the formula, but have a look at the graph in figure 6.

On this estimate, the fertility caused by Y-chromosome decay drops to 1 per cent of its present level within 5,000 generations, which is about 125,000 years. Not exactly the day after tomorrow – but equally, not an unimaginably long time ahead. Very roughly, in fact, as long into the future as our species has been going so far from its beginnings in Africa. Some other factors might work either to extend this time or to reduce it. For example, there might be very robust Y-chromosomes around that are less vulnerable to mutational attack and could take over from others as they are

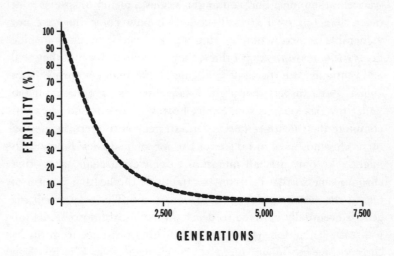

Figure 6: The extinction of men: decay of the Y-chromosome

eliminated one by one. The chromosome's newly discovered ability for internal recombination might help slow down the process of decay and put back the time to eventual extinction. Or, on the other hand, the proportion of mitochondrial mutations which severely cripple sperm could increase with time; this will accelerate the process. Or the assumptions I made at the beginning about the rate of wounding mutations and their effects on fertility might be inaccurate, making the estimate either too long or too short. But whatever these modifications affecting the timescale, the continuous and cumulative decay of the Y-chromosome will progressively and inexorably reduce male fertility to the point where men do become extinct. I deliberately use 'men' instead of 'our species' because only men require a Y-chromosome. Of course, unless something changes in the way we breed, women will vanish too and our entire species *Homo sapiens* will disappear at some time in the next one or two hundred thousand years. But is that inevitable?

Extinctions happen all the time. It is expected and it occurs and we are not immune. But, you might well ask, plenty of species a lot older than our own are still going, so how come they are not vulnerable to extinction by the same process of chromosome decay? My answer is that I think they are vulnerable and they will all eventually face the same challenge. I am unable to prove it, of course, but I suspect that a good many species have already gone under for this very reason. Some, however, have found a way to commute their death sentence. One strategy is to recruit genes on other chromosomes to take over the job of male development. It might take only a small mutation to convert a gene on another chromosome so that it becomes capable of duplicating the job of one of the endangered Y-chromosome genes. This way, when the gene is eventually battered to death on the Y-chromosome, its job is already being done elsewhere and its disappearance from the Y-chromosome no longer matters. An element of luck is involved since, as nothing is planned in evolution, if this rescuing mutation had not happened in at least one individual, the species would have disappeared – as many must have done. So many genes concerned with male development are found on other chromosomes that it is a safe bet that many of them have already escaped from the decaying Y-chromosome before it became too late. But it is a very risky strategy, and failure to have a back-up somewhere else will have driven many species to extinction as their Y-chromosomes decayed away.

It is a race against time. Can a species get the genes it needs off the Y-chromosomes, or recreate them elsewhere, before it goes under? Always the last gene to go will be SRY, the master switch itself. We know it is capable of jumping ship and smuggling itself onto another chromosome. The evidence for this abandonment lies in the few cases of XX males. To be male at all, they must have an SRY gene to start the ball rolling towards male development, but in these men SRY cannot be located on the Y-chromosome – because they don't have one. Their SRY genes are stowed away on another chromosome. In the germline cells of their fathers, SRY became detached from the Y-chromosome and re-implanted somewhere else. However, in its hurry to leave the Y-chromosome in these men, SRY has left the other genes behind. The few remaining genes on the Y-chromosome are still necessary for proper sperm production, so the XX men are, sadly, sterile. If these few genes had already left the Y-chromosome, or had their functions been reproduced elsewhere, the XX men would have been completely fertile and the species would be saved. Everything needed to make a man would then have escaped the doomed Y-chromosome and it could have been left to rot.

In a variation on this theme, another possibility open to the inventive is to bypass SRY altogether by switching on the male development process a step or two down the chain of command. These secondary relays, the genes switched on by the signal from SRY, are already safely stowed on other chromosomes. A lucky mutation in one of these could activate the relay without waiting to get the nod from SRY. No longer needed, SRY could be left to its fate.

All of these ways of escaping from the dying Y-chromosome are risky and need a lot of preparation, for instance in relocating the sperm production genes before finally jumping ship. Lots of species will have tried this to avoid extinction, but it seemed that none had succeeded. Then, in 1995, researchers found a mammal which had managed to escape this fate. When they looked under the microscope at the chromosomes of a small burrowing rodent called the mole vole, *Ellobius lutescens*, which lives in the foothills of the Caucasus mountains, they discovered that the males didn't have a Y-chromosome. Neither, it turned out, did they have an SRY gene. This inconspicuous little rodent had managed to complete that last manoeuvre and activate a gene relay one or two stages down the line from SRY. And only just in time. The mole vole Y-chromosome has now completely disappeared. The mole vole is now safe from Y-chromosome-driven extinction, the only mammal species known to have succeeded in getting itself out of danger. The new mole vole master switch, whatever it is, will over time convert the chromosome it is on into a lonely outcast, just as surely as the Y-chromosome was doomed to a slow and humiliating decline as soon as SRY took up the baton of directing male development. For the mole vole, the problem has been shelved for tens of millions of years. For us and all other mammals who still have to rely on a Y-chromosome to make males, the danger is much more immediate.

One thing distinguishes our species from the others that face extinction through their reliance on a rapidly disintegrating chromosome. Unlike the rest, we are at least capable of being aware of our impending doom. The mole vole has no idea how close he came to extinction or why, by relocating the male master switch, he was able to avoid it. Nor did the far greater number of mammal species which left it too late to abandon the crumbling Y-chromosome realize what was going on as it destroyed them. Only our species, in the whole history of the planet, has the knowledge and the capability of understanding and perhaps even averting this otherwise certain fate. The questions we face boil down to this. Do we need men? Can we do without them? And if we can be bothered, what should be done to save them?

Many people would rejoice at the extinction of men. Valerie Solonas was one. She is best known as the woman who shot Andy Warhol in 1968. The previous year she published the venomous SCUM manifesto, which begins:

Life in this society being, at best, an utter bore and no aspect of society being at all relevant to women, there remains to civic-minded, responsible, thrill-seeking females only to overthrow the government, eliminate the money system, institute complete automation and destroy the male sex.

The expanded acronym of her manifesto title – the Society for Cutting Up Men – leaves us in no doubt as to Ms Solonas's preferred solution to the world's problems, but unless other arrangements are put in place, their demise will take women with them. Destroying the male sex might get rid of men, but it would be a very short-lived victory. Men are still required for breeding, if nothing else. As things stand just now, sperm are needed. But for how much longer?

The wide application of ICSI, the fertilization of eggs by injecting sperm, could delay the extinction by allowing men to breed whose Y-chromosomes have rotted away so much that they are no longer capable of producing viable sperm. Yet even if ICSI became the norm in some future century, it would still not prevent the progressive deterioration of the Y-chromosome. In fact it would accelerate it, by saving terminally sick chromosomes from being weeded out when, in the normal course of events, they would disappear by producing infertile sons. In time, as the inexorable decay continued, men would become increasingly dependent on ICSI until there would be no Y-chromosome remaining anywhere which was sufficiently intact for the man who carried it to be able to breed without help. ICSI can delay the extinction of men, but will not prevent it. The SRY gene itself is not immune to decay and its demise will be terminal. ICSI can prolong the life of Y-chromosomes no longer capable of making sperm that work properly, but it cannot rescue Y-chromosomes that are no longer able to make men. When that gene is hit, the Y-chromosome gets through to the next generation but it no longer has the power to manufacture males; the children who inherit it will be XY females, incapable of breeding, even with the help of ICSI. They are women and produce no sperm to inject.

Although ICSI will not prevent the extinction of men, it is at least a technique which we know works. The other remedies that spring to mind have yet to be proved effective, but if men are to be retained they are at least worth considering. For instance, what would happen if we deliberately abandoned the Y-chromosome and switched the necessary genes to the other chromosomes where they would be safe? In other words, if we pre-empted the demise of the Y-chromosome and deliberately engineered the solution so fortuitously arrived at by the mole vole? The human Y-chromosome could be left to decay – it cannot be saved – but men would be reprieved. But could this be made to work? We now know, thanks to David Page's work, all the genes that are present and necessary on the Y-chromosome to make a man in full working order. Even with today's comparatively primitive genetic engineering technology it would be comparatively easy to cut them out of the wreckage of the Y-chromosome and assemble them together in a compact genetic package. Or they could be made from scratch, even with present-day DNA synthesis instruments. From there, it would be a relatively straightforward task to insert the package into another chromosome, and the chances are it would work straight away. We saw in an earlier chapter how a fertilized mouse egg destined to become female had been successfully diverted to at least superficial masculinity by the injection of the mouse equivalent of SRY. Sure, it was infertile; but if the egg had been injected with the complete package of male genes, the mouse would have been both male *and* fertile.

A fertilized human egg which would otherwise develop into a girl would, given this treatment, grow into a perfectly healthy man indistinguishable from any other, until you looked at his chromosomes. He would have two X-chromosomes; but, instead of being infertile like XX males today, this man would have all the necessary sperm genes. But what about his own children? No immediate problem there either. Assuming that the package of male genes had landed safely on one chromosome, this new-age Adonis would be able to have sons and daughters in equal proportion, their sexual destinies decided only by whether they received from him a sperm carrying the repackaged chromosome with the added genes (for sons) or an original (for daughters). From then on it would be plain sailing. The Adonis chromosome would carry on untroubled by the deteriorating condition of the Y-chromosomes in other men. It would never meet one because, after all, it is not going to be fertilizing other sperm, only eggs. The prospects for the Adonis chromosome are excellent. It will reprieve men from the brink of extinction and guarantee them a future for several million years. There is no genetic reason why several different versions of the Adonis chromosome could not be circulating at the same time. After all, when these chromosomes are created by injection of the male gene package those genes could, with present technology, land almost anywhere. Assuming that many men are created in this manner, each with the manly gene package posted to a different chromosomal address, there would be a profusion of different Adonis chromosomes. But even that would not matter because, for the same reason that an Adonis chromosome will never meet a Y-chromosome, there is no danger of breeding a man with more than one Adonis chromosome. A man can only mate with women, who cannot have an Adonis chromosome – or they would be men too. The Adonis chromosome looks a good bet to me. I almost wish I had one myself.

The purpose of all this effort and ingenuity is to avoid the otherwise inevitable eventual extinction of men, and with them our entire species. We have seen how the widespread use of ICSI will lend a crutch to progressively crippled Y-chromosomes. We have dreamed up a genetic engineering solution by creating a new range of Adonis chromosomes with excellent long-term prospects that keeps men going but abandons the Y-chromosome to its fate. But while the chromosome may have gone, the ingredients of Adam's Curse are still there: two sexes, sperm and eggs, and sexual selection with all its consequences. Gaia would go on suffering. One final genetic solution that I offer for scrutiny is the most radical. That is to abandon men altogether. It sounds impossible but, from the genetic point of view, very little stands in its way – and it would lift the Curse once and for all. Consider what is happening when sperm meets egg. The sperm brings with it a set of nuclear chromosomes from the father which, after fertilization, mixes in with a set of nuclear chromosomes from the mother. What is to stop the nuclear chromosomes coming not from a sperm but from another egg?

Let's think this through a little more. We know from ICSI that sperm can be injected into eggs. If we can do that, there is nothing to stop the nucleus from a second egg being injected instead. That would be very easy. But would it develop normally? At the moment the answer is no, but it is short-sighted to say that it is fundamentally impossible. The snag, and I would not call it anything stronger, is that during the time they spend in the two different germline cells, male and female, the chromosomes are censored. It is a process called *imprinting* and – very briefly – means that about fifty genes on several different chromosomes are crossed out with the genetic equivalent of a censor's blue pencil. No-one is quite sure why this happens, though many believe it to be yet another aspect of the war between the sexes. Without going into any of the details, the fact is that female and male germline cells cross out different genes. The fertilized egg, and the somatic cells of the embryos that develop from it, cannot read what lies behind the censor's mark. Normally that doesn't matter. Because each parent crosses out different genes and all the cells have one chromosome from each parent, the cells can take their instructions from the uncensored copy. If gene A is censored in the egg, it will be readable in the sperm. Likewise, gene B, crossed out in the sperm, can be read from the egg's copy. The snag is that if *both* sets of chromosomes come from the egg, as in the male extinction scheme we are presently evaluating, the censor's pencil will have struck out the same gene on both copies and our embryo will not have its full set of instructions in legible form.

The reason I don't see this as more than a temporary nuisance is that we know there are natural systems which are capable of erasing the crossings-out. When the chromosomes from the developing embryo are processed through its own germ line ready to be passed on to the next generation, all the pencil marks are rubbed out. Only then do the germline cells start crossing out genes, their choice of text for censorship depending on whether they are making eggs or sperm. I don't know exactly how this hurdle can be crossed, but it seems to me to be far from insuperable. There is nothing else, genetically speaking, to stand in the way. The young embryos growing from the female x female fertilized eggs would re-implant as easily as any other following regular IVF treatment. There they would grow into a perfectly normal foetus and be born as a perfectly normal baby. The only difference from any other birth is that the sex is always predictable. The baby is always going to be a little girl. The entire process has been accomplished without sperm, without Y-chromosomes and without men.

Importantly, the baby girls will not be clones. The media coverage following the birth of Dolly the sheep looked forward with hope or revulsion, but usually the latter, to the birth of the first human clone. In late 2002, an obscure cult, the Raelians, announced the birth of a cloned girl in Canada, though it was almost certainly a hoax. An Italian doctor, Severino Antinori, has made a habit of announcing the imminent birth of human clones – though none has yet arrived. On top of the moral and ethical objections to cloning, it is not a successful long-term strategy for the entire species, for reasons we have covered in an earlier chapter. As they are denied the genetic advantages of recombination, clones are very susceptible to their mother's parasites.

But our baby girls are not clones. They are the same mixture of their parent's genes, shuffled by recombination just as thoroughly as any of today's children. They have two biological parents, not just one. Their only difference from any other child is that both parents are women. Instead of a father and mother, these girls have two biological mothers. From a genetic point of view, they are completely normal, indistinguishable from any little girls around today. In a world with men, when the girls grew up they would be able to breed in the old-fashioned way just as easily as women today. With all these advantages, and assuming the low hurdle of imprinting can be cleared, I am quite sure that someone will try this before very long. Lesbian couples already enlist the help of a man to donate his set of chromosomes to fertilize the eggs of one of them. How much more attractive for these couples to have a baby to whom both, rather than just one of them, were parents. It is bound to happen. Men are now on notice.

But would it catch on, and could it be a solution to the extinction of our species posed by the crumbling Y-chromosome? That is harder to say. Once men entirely disappeared, and were long forgotten, all reproduction would need to be assisted to some extent. There is no purely genetic objection that I can see, but the prospect brings with it a host of other issues. Certainly if the wholesale extinction of men were to be purposefully and deliberately engineered, whether by methods akin to Valerie Solonas' direct action proposal or by more devious means, this Sapphic form of reproduction would have to be in place before men were dispensed with altogether. The Solonas solution of wholesale slaughter is messy and, while I was wondering about alternative possibilities, I heard that a Belgian biologist and journalist, Dirk Draulans, had devised a scheme that would be hard to beat. In his novel *The Red Queen* Draulans also anticipated the possibility of egg–egg fertilizations. I won't spoil a good story by telling you the ending, but the way the men are seen off is brilliantly subtle. A virulent virus is genetically engineered to lock onto the Y-chromosome, which would not really be difficult to achieve. But that is not the subtle part. It would be no good giving all men the same viral disease; someone would notice and take steps to contain the epidemic. The beauty of Draulans' plan is that the virus, now stuck to the Y-chromosome of all men who catch it – and it is naturally extremely virulent – then begins to make an enzyme which imitates the process of X-inactivation that takes place in female cells and shuts down the X-chromosome that shares every male cell. Men and boys fall prey to a whole range of sicknesses, each caused by inactivation of the male X-chromosome. So, gentlemen, be warned: the plan for your imminent extermination is already hatched, if only in a novel.

I will leave you to imagine a world without men, but there is one immediate benefit from their extinction. Adam's Curse is permanently lifted. Sexual selection disappears, for the simplest of reasons – there are no longer two sexes. Sperm no longer fights sperm for access to eggs. There are no sperm to do battle, no Y-chromosomes to enslave the feminine. The destructive spiral of greed and ambition fuelled by sexual selection diminishes and, as a direct result, the sickness of our planet eases. The world no longer reverberates to the sound of men's clashing antlers and the grim repercussions of private and public warfare. The great sexual experiment, begun eons ago in our single-celled ancestors, is over. Mitochondria and the female have finally triumphed over their ancient adversaries . . . and Gaia can resume her broken sleep.

AFTERWORD

When I wrote the final chapter, and looked into the future, I predicted that the hurdle of genetic imprinting, which prevents both parthenogenesis and successful female x female fertilizations in mammals, would one day be overcome. I never imagined that it would have been accomplished within weeks of the publication of *Adam's Curse*. On April 22, 2004, the journal *Nature* carried a report from a group of Japanese and Korean scientists of a mouse born without a father, created by the fertilization of the egg from one female mouse with the egg of another. They had succeeded in overcoming the barrier of genetic imprinting by removing just one imprinted gene from each parent, the unexpected effect of which was to overcome all the other imprinted genes. Instead of dying in the womb, the embryo developed normally and was born alive and well nearly three weeks later. Even more amazingly, the mouse (she was called Naguya after a young girl in a Japanese fairy tale found abandoned in a clump of bamboo) grew into a normal adult and, in classic fashion with a male, became a mother with a litter of her own.

Several hundred attempts at egg x egg fertilization by the researchers failed, and Naguya was the only mouse to survive to adulthood, so this was by no means an easy procedure. But it did show that the hurdle of genetic imprinting was much lower than I and most other scientists had imagined. The Japanese team is planning to try the same technique on other mammals and, though no one seriously proposes attempting this with humans in the immediate future, it does bring the prospect of girls being born with two mothers and no father a good deal closer.

—Bryan Sykes
Oxford
January 2005

INDEX

ABOUT THE AUTHOR

Bryan Sykes, Professor of Human Genetics at the University of Oxford, is one of the world's leading geneticists. After undertaking medical research into the causes of inherited bone disease, he discovered that DNA could survive in ancient bones, and he was the first to report on the recovery of ancient DNA from archaeological bones in the journal *Nature* in 1989. Since then Professor Sykes has been called in as the leading international authority to examine several high-profile cases, such as the Ice Man, Cheddar Man and the many individuals claiming to be surviving members of the Russian royal family. As well as a scientist, Professor Sykes has been a television news reporter and a parliamentary science adviser.

The publication of Dr. Sykes's first book, *The Seven Daughters of Eve*, resulted not only in a *New York Times* bestseller but also in a work that received widespread international praise. The *Wall Street Journal* called it "a lovely, rollicking book, direct and clear. . . . [A] fascinating glimpse into anthropology in the era of the genome." *Nature* raved, "Scientifically accurate and understandable to the layperson. . . . [*The Seven Daughters of Eve*] will be recognized as an important work, bringing molecular anthropology to a mass audience." The UK's *Evening Standard* wrote, "A signal success, effortlessly bringing the reader up to its breezy speed, with fascinating case histories ranging from the remains of the Russian royal family to pet golden hamsters."

About Oxford Ancestors

Dr. Sykes and his research team have, over the last ten years, compiled the most complete DNA family tree of our species yet seen. In 2001, following the publication of *The Seven Daughters of Eve*, Professor Sykes was inundated with demands from people who wanted to know more about their own ancestry and founded Oxford Ancestors Ltd, which has become the world's leading provider of DNA-based services for use in personal ancestry research. Find out more about Oxford Ancestors and the services it provides by visiting their Web site: http://www.oxfordancestors.com.